Camila de Souza Varize
Augusto Bücker
Luiz Carlos Basso

Ethanol fermentation

Camila de Souza Varize
Augusto Bücker
Luiz Carlos Basso

Ethanol fermentation

Modified strains and amino acid supplementation

ScienciaScripts

Imprint

Any brand names and product names mentioned in this book are subject to trademark, brand or patent protection and are trademarks or registered trademarks of their respective holders. The use of brand names, product names, common names, trade names, product descriptions etc. even without a particular marking in this work is in no way to be construed to mean that such names may be regarded as unrestricted in respect of trademark and brand protection legislation and could thus be used by anyone.

Cover image: www.ingimage.com

This book is a translation from the original published under ISBN 978-613-9-62053-1.

Publisher:
Sciencia Scripts
is a trademark of
Dodo Books Indian Ocean Ltd. and OmniScriptum S.R.L publishing group

120 High Road, East Finchley, London, N2 9ED, United Kingdom
Str. Armeneasca 28/1, office 1, Chisinau MD-2012, Republic of Moldova, Europe
Printed at: see last page
ISBN: 978-620-7-71200-7

SUMMARY

ACKNOWLEDGMENTS

First of all, I thank God for the blessings he has given me;

To my fiancé, Lucas Dantas Lopes, for his immense love and companionship;

To my mother, Eliane de S. Varize, for all her love and for being my main source of support;

To my sisters Angélica and Caroline de S. Varize and my father Edinaldo Varize (*in memoriam*), for all their love and affection;

To the rest of my family and my fiancé's family, for the affection and good times we shared;

To my advisor, Prof. Dr. Luiz Carlos Basso, for all his scientific teachings, for always being there for me and for the good interaction;

To my co-supervisor, Prof. Dr. Augusto Bücker, for always being helpful and for his contributions to the improvement of this work;

To Prof. Dr. Boris U. Stambuk, for providing the yeast strains;

To the laboratory technician, Luis Lucatti (Cometa), for his friendship and help on the bench;

To Prof. Dr. Luiz Humberto Gomes (Beto), for his attentive communication and for making the laboratory available whenever I needed it;

To my friends in the lab, Renata, Mariane, Carolina, Cristiane, Osni, Ricardo, Thalita and all the other colleagues who have worked in the Yeast Biochemistry and Technology Laboratory at ESALQ/USP, for their help and good fellowship. In particular, Renata, Mariane and Carolina, for their friendship, and because they were present until the last moment of bench testing, actively helping to make this work a reality;

To the post-graduate program in Agricultural Microbiology at ESALQ/USP;

To the CNPq agency, for the grant during the PhD period.

Camila Varize

Men are like rivers

One of the most ingrained and generally accepted precepts is the belief that men have immutable qualities: there are good or bad men, intelligent or stupid, energetic or apathetic, and so on. Well, men aren't like that. We can just say that a man is more often good than bad, more often intelligent than stupid, more often energetic than apathetic, or the opposite; but to classify one man, as we always do, as good or intelligent and another as bad or stupid is a mistake. Rivers too, all rivers of water, are sometimes narrow, sometimes fast, sometimes wide or calm, transparent or cold, flowing or slow. Men are like rivers. Each one carries with it the seed of all human qualities, revealing some characteristics at certain stages, others at others, and even, on certain occasions, showing itself in a form completely opposite to its inner nature, which it nevertheless retains.

(in Ressiirreiccio)

Leon Tolstoy

SUMMARY

Ethanol fermentation: modified strains and amino acid supplementation

Increasing the share of biofuels in the world's energy mix could help to prolong the existence of oil reserves, mitigate the threats posed by climate change and improve the security of energy supplies on a global scale. In this scenario, the Brazilian process of producing ethanol from sugar cane has gained prominence due to its high yields and low production costs. *S. cerevisiae* strains are widely used in industrial fermentations and, although they are considered more tolerant than others, the Brazilian process imposes a variety of stress factors on them, affecting their metabolism and growth. High-alcohol fermentation, carried out using musts containing high concentrations of sugars, is one of the most efficient ways of obtaining high levels of ethanol. However, this technology has additional deleterious effects on the yeast. In this context, increasing yeast tolerance is of fundamental importance in order to achieve satisfactory fermentation performance. This study evaluated *S. cerevisiae* strains, isogenic to the CAT-1 industrial strain, with over-expression of the *TRP1* and *MSN2* genes, involved in tryptophan biosynthesis and the general response to stress, respectively. These strains were evaluated for their potential to carry out high-alcohol fermentations, simulating Brazilian industrial conditions. The results showed that the *MSN2* gene, in its truncated version, favored the strain mainly in relation to osmotic stress, increasing fermentation speed and sugar consumption. The *TRP1* gene promoted greater growth of the strain in YEPD medium with 8% ethanol, but made the strain less viable at concentrations above this level. This study also assessed the effect of amino acid supplementation on the physiology of the CAT-1 strain in YNB medium and in molasses and sugar cane syrup musts. Histidine supplementation promoted greater cell growth and viability in the different media tested. In addition to histidine, the amino acids lysine and alanine increased CAT-1 growth in molasses must. Tryptophan and asparagine supplementation also increased cell viability in syrup mash. On the other hand, in microplate tests, supplementation with cisterna degraded the growth of the strain in

YNB medium with 10 and 12% ethanol and in molasses mash with 20% ART. The results obtained indicate that both genetic engineering and amino acid supplementation could be viable alternatives for increasing the tolerance of *S. cerevisiae* to withstand the multiple stress conditions found in Brazilian distilleries.

Keywords: *S. cerevisiae*; Osmotic stress; High alcohol content; Amino acids in fermentation; Alcoholic fermentation

1 INTRODUCTION

Due to global warming and the future depletion of oil reserves, the importance attached to biofuels has increased enormously. The positive impacts of using biofuels to replace fossil fuels are not restricted to the economic sphere, but also involve political and environmental issues. In addition to reducing dependence on oil and energy costs, the use of biofuels can result in a significant reduction in greenhouse gas emissions. In this scenario, the Brazilian sugar-alcohol sector has stood out due to the high yield and low cost of producing fuel ethanol, with sugar cane being the main raw material used.

In Brazil and the USA, ethanol production technologies have already been consolidated, which makes this biofuel a strong candidate to participate in the world's energy matrix on a large scale. Ethanol is the most widely used biofuel in the world, both in pure form and as an additive to gasoline (Mckendry, 2002; Sanchez et al., 2008).

In the Brazilian process, sugarcane juice or molasses is used as a substrate for the production of ethanol by the yeast *Saccharomyces cerevisiae*. During industrial fermentation, various stress factors are imposed on the yeast, promoting deleterious effects on it, affecting its metabolism and growth. These factors include: high sugar concentration, the presence of salts in the must, high temperatures, high ethanol concentration, low extracellular pH, bacterial contamination, nutrient deficiency and the presence of inhibitory agents in the must. In addition, in this process the yeast biomass is reused in several fermentations each day, which can contribute to a significant decrease in cell viability (Basso et al., 2011).

Technologies have been used to increase the fermentation efficiency of ethanol production. High-alcohol fermentation, known as *Very High Gravity Fermentation,* for example, is one of the ways of obtaining higher ethanol yields from fermentation of musts containing more than 25% (w/v) total sugars (Thomas et al., 1994). Although the yeast *S. cerevisiae* is widely used in the process and is considered to be more resistant to various types of stress, research efforts are still needed in order to make progress in terms of its tolerance to the industrial fermentation process. Studies using genetic engineering of yeasts have aimed to increase the resistance of the cells to withstand deleterious conditions. According to Bücker et al. (2014), modifying the promoter region of the *TRP1* gene, involved in tryptophan biosynthesis, and the *MSN2* gene, which encodes a transcription factor (Msn2) involved in the general stress response in *S. cerevisiae,* can increase the tolerance of S. *cerevisiae* strains to high alcohol content and other types of stress faced in the Brazilian process. It has also been proposed that the use of amino acids to supplement the fermentation must can promote greater tolerance and growth of yeast cells in the face of different stress conditions, such as high alcohol content and osmotic stress caused by high sugar concentrations in the medium (Thomas & Ingledew, 1990; Pham & Wright, 2008; Hu et al., 2005; Blomqvist et al., 2012). In addition to performing vital functions for cells, such as synthesizing enzymes and structural components, amino acids can also be incorporated into cytoplasmic membrane proteins, promoting greater membrane rigidity and cell tolerance (Hu et al., 2005).

This study was divided into two parts. In the first, the aim was to evaluate the potential of *S. cerevisiae* strains, isogenic to the CAT-1 industrial strain, with over-expression of the *MSN2* or *TRP1* genes, for fermentations with high alcohol content and cell recycling, simulating the industrial conditions of Brazilian distilleries. In the second, the aim was to evaluate the effect of amino acid supplementation on the growth, cell viability and ethanol production of CAT-1 yeast in fermentations of molasses must and sugar cane syrup, using cell recycling.

References

Mckendry, P. Energy production from biomass (part 2): conversion technologies. **Bioresource Technology**, v. 83, n. 1, p. 47-54, 2002.

Sânchez, O.J.; Cardona, C.A. Trends in biotechnological production of fuel ethanol from different feedstocks. **Bioresource Technology**, v. 99, n. 13, p. 5270-5295, 2008.

Basso, L.C.; Basso, T.O.; Rocha, S.N. Ethanol production in Brazil: the industrial process and its impact on yeast fermentation. In: SANTOS BERNARDES, M.A. (Ed.). **Biofuel Production - Recent Developments and Prospects**. InTech, 2011, chap. 5, p. 85-100.

Thomas, K.C.; Hynes, S.H.; Ingledew, W.M. Practical and theoretical considerations in the production of high concentration of alcohol by fermentation. **Process Biochemistry**, v. 31, n. 4, p. 321-331, 1996.

Thomas, K.C.; Ingledew, W.M. Fuel alcohol production: effects of free amino nitrogen on fermentation of very-high-gravity wheat mashes. **Applied and Environmental Microbiology**, v. 57, n. 7, p. 2046-2050, 1990.

Pham, T.K.; Wright, T.K. The proteomic response of *Saccharomyces cerevisiae* in very high glucose conditions with amino acid supplementation. **Journal of Proteome Research**, n. 7, p. 4766-4774, 2008.

Blomqvist, J.; Nogué, V.S.; Gorwa-Grauslund, M.; Passoth, V. Physiological requirements for growth and competitiveness of *Dekkera bruxellensis* under oxygen-limited or anaerobic conditions. **Yeast**, v. 29, n. 7, p. 265-274, 2012.

Hu, C.K.; Bai, F.W.; An, L.J. Protein amino acid composition of plasma membranes affects membrane fluidity and thereby ethanol tolerance in a self-flocculating fusant of *Schizosaccharomyces pombe* and *Saccharomyces cerevisiae*. **Sheng Wu Gong Cheng Xue Bao**, v. 21, n. 5, p. 809-813, 2005.

Bücker, A. **Genomic engineering of an industrial strain of Saccharomyces cerevisiae to improve ethanol tolerance**. 2014. 111p. Thesis (PhD in Biochemistry) - Federal University of Santa Catarina, Florianópolis, 2014.

2 literature review

2.1 The importance of bioethanol

Petroleum-derived fuels, such as diesel oil and gasoline, are widely used around the world. However, there is a recurring concern that this natural resource is finite (Demirbas & Balat, 2006; Searchinger et al, 2008). In addition, the burning of fossil fuels is responsible for the emission of a significant proportion of greenhouse gases (GHG). In this scenario, carbon dioxide (CO_2) is the most produced gas, contributing to climate change (Wuebbles & Jain, 2001; Gallo, 2011).

The use of renewable energy is one of the most efficient ways of achieving sustainable development. Increasing its share in the world's energy mix would help prolong the existence of fossil fuel reserves, mitigate the threats posed by climate change, and allow for better security of energy supply on a global scale (Mckendry, 2002; Gray et al., 2006; Goldemberg, 2007; Nigam; Singh, 2011). Most "new renewable energy sources" are still in the commercial development phase, but some technologies are already well established, such as Brazilian sugarcane ethanol, which after 40 years of production has become fully competitive with gasoline and suitable for use in many countries, becoming a global energy commodity (Goldemberg, 2007).

According to Balat et al. (2009), ethanol is the most widely used biofuel in the world and can be produced from different types of raw materials, classified into three categories: simple sugars, starch and lignocellulose. It can also be used intensively in the chemical, beverage, pharmaceutical and cleaning industries (Penido-Filho, 1980).

Brazil and the United States are the largest producers of ethanol in the world (Gerbens-Leenes et al., 2012), with production in the USA being greater and based on the use of corn, while in Brazil sugar cane is used almost exclusively (Wheals et al., 1999; Walter et al., 2008; Gerbens-Leenes et al., 2012). In 2016, a total of 100.7 million m3 of fuel ethanol was produced worldwide. In the same year, the USA and Brazil accounted for 57.6% and 27.4% of the world's total ethanol production, respectively (RFA, 2017).

The use of sugar cane to produce ethanol, as well as sugar, was a political and economic decision that involved investment by the Brazilian government in 1975, when the Federal Government decided to encourage the production of alcohol as a substitute for gasoline, with the idea of reducing oil imports due to the crisis faced. The emergence of the National Alcohol Program (Proàlcool) presented positive environmental, economic and social aspects, and became the most important biomass energy program in the world (Goldemberg et al., 2004).

Sugarcane, in the form of juice or molasses, can be used in tropical countries to produce ethanol. This raw material is mainly composed of sucrose, fiber and water, usually in proportions of approximately 12%, 15% and 70% respectively. The other components are other sugars (glucose and fructose), inorganic materials, nitrogenous substances, gums, waxes and organic acids (Chen & Chou, 1993).

The conversion of sucrose, contained in this plant, into ethanol is easy compared to starch materials and lignocellulose, because prior hydrolysis of the raw material is not necessary, since this disaccharide can be directly metabolized by yeast cells, mainly belonging to the *Saccharomyces cerevisiae* species (Waclawovsky et al., 2010). In *S. cerevisiae* yeasts, sucrose is metabolized mainly via the extracellular pathway (Gascón & Lampen, 1968). In this pathway, an enzyme naturally produced by this yeast species, called invertase (β-D-fructosidase), hydrolyzes sucrose in the periplasmic space, i.e. in the extracellular space, forming the monosaccharides glucose and fructose. These monosaccharides are transported to the intracellular space and subsequently metabolized in the glycolytic pathway (Barnett, 1976; Barnett, 1981; Walker, 1998; Lagunas, 1993). Ethanol produced from sugar cane not only has a positive energy balance, but has also benefited from the support of government policies in several countries, including Brazil, which supplies approximately 40% of this biofuel to passenger vehicles, i.e. a third of its total energy demand for transportation (Goldemberg, 2009).

Due to increasing environmental concerns and periodic crises in oil-exporting countries, ethanol has become a viable and realistic alternative in the world energy market. Therefore, the development of low-cost technologies for the production of fuel ethanol is a priority for many research centers, universities, private companies, and even different public policies (Cardona & Sanchez, 2007). The need to increase ethanol production in order to serve a potential market, combined with dependence on fossil fuels, poses challenges for the Brazilian government (Duarte et al., 2013). Knowing that the supply of fossil fuels will not be able to meet our planet's growing energy needs in a sustainable way, the market conditions related to supply and demand analyzed make ethanol production promising (Cerqueira-Leite et al., 2009).

In view of this reality, the increase in the production of renewable energies will grow progressively. Environmental, economic and political issues will be key to sustaining the growth of ethanol production (Gonçalves et al., 2011). Several obstacles in the alcoholic fermentation process must be overcome in order to achieve a higher and more competitive performance in ethanol production by the yeast *Saccharomyces cerevisiae* (Alfenore et al., 2002; Guardabassi & Goldemberg, 2014).

2.2 Alcoholic fermentation and the yeast *Saccharomyces cerevisiae*

Alcoholic fermentation is a biological process in which ethanol is produced through the action of yeasts, mainly belonging to the *S. cerevisiae* species, in media containing fermentable sugars (Bai et al., 2008).

The yeast *Saccharomyces cerevisiae* is a unicellular, eukaryotic, heterotrophic fungus belonging to the phylum Ascomycota. Its reproduction is usually asexual, more specifically by budding. It has a high degree of cellular organization, containing a cell nucleus with a nuclear envelope and organelles (Madigan et al., 2010).

As well as being used to produce ethanol, *S. cerevisiae* can be used in the pharmaceutical, bakery and beverage industries. It is also known that it is widely used as a model organism in studies of the biochemistry, genetics and cell biology of eukaryotes. Its biological knowledge is well developed and it is considered easy to maintain and manipulate in laboratories (Ostergaard et al., 2000; Zakrajsek et al., 2011).

In the ethanol production process, *S. cerevisiae* yeasts have several desirable and advantageous characteristics when compared to other yeast species. Examples of these attributes are the high and rapid ability to transform fermentable sugars into ethanol, the ability to survive in environments with higher acidity, osmotolerance, the ability to tolerate high ethanol contents and the ability to withstand temperature variations (Andrietta et al., 2007).

The metabolic pathway involved in alcoholic fermentation is glycolysis. In this pathway, one molecule of glucose is metabolized and two molecules of pyruvate are produced. The molecules of pyruvic acid are converted into ethanol, with the release of 2 molecules of CO_2 and the formation of 2 molecules of adenosine triphosphate (ATP) (Walker, 1998). Alcoholic fermentation takes place in the cytoplasm of yeast cells, through glycolysis followed by the reduction of pyruvate, where after 12 enzymatic reactions, ethanol and CO_2 are produced. The total oxidation of sugar by the respiratory chain takes place in the mitochondria of the cells, in which case the energy balance is higher (Castrillo & Ugalde, 1994; Lagunas, 1981).

Yeasts, like all other microorganisms, have as their main objective the perpetuation of the species, ethanol being a waste product excreted by them in anaerobiosis (Lagunas, 1976). An evolutionary explanation for the preference of the fermentative pathway - which results in a lower energy yield - over the respiratory pathway under certain conditions, is that ethanol is excreted by the cells to avoid a scramble for nutrients. This phenomenon is called an antagonistic relationship, since ethanol is toxic to many other microorganisms (Golubev, 2006). *Saccharomyces cerevisiae* yeasts are conceptualized as Crabtree positive, because even in the presence of oxygen, they carry out fermentation when glucose or other fermentable sugars are available in the medium (Pronk et al., 1996; Lagunas, 1981).

According to Lagunas (1976), the primary objective of yeasts, when in anaerobiosis, is to use the carbon source to generate a form of chemical energy (ATP), energy that will be used to carry out the biosynthesis of molecules and also to carry out physiological work, both necessary for the maintenance of life: growth, multiplication and perpetuation of the species. During alcoholic fermentation, part of the carbon source is diverted to the synthesis of proteins, nucleic acids, polysaccharides and lipids necessary for the production of biomass. Other products related to adaptation and survival are also formed, such as glycerol and organic acids (Reed & Nagodawithana, 1990; Amorim et al., 1996). Glycerol, organic acids (succinic and acetic acids) and biomass are by-products related to the oxidation-reduction equilibrium in anaerobiosis. Alcoholic fermentation is a non-oxidative process and in order to maintain the oxidation-reduction balance, all the NADH formed in oxidation reactions resulting from the production of biomass and the formation of organic acids must be consumed in reduction reactions, coupled with the production of ethanol and glycerol (Blomberg & Adler, 1992; Van-Dijken & Scheffers, 1986). Glycerol is also an osmoregulatory metabolite and is the secondary product formed in the greatest quantity by yeasts, originating from the same pathway in which ethanol is synthesized. The formation of glycerol is increased when there is high osmotic pressure in the medium, protecting the cells in the presence of this stress (Guo et al., 2001; Cronwright et al., 2002).

2.3 Brazilian industrial process and stressful conditions for yeasts

Alcoholic fermentation is a process that depends on several factors, including the species of yeast used, temperature, oxygen supply rate, pH and the chemical nature of the nutrients supplied (Ceccato-Antonini, 2008; Querol et al., 2003).

In Brazil, the must that is normally used for industrial fermentations comes from molasses resulting from the final crystallization of sugar, and its composition includes: water, fermentable carbohydrates (glucose, fructose and sucrose), compounds of organic origin (amino acids, vitamins, proteins) and a mineral fraction (composed mainly of calcium, magnesium, sodium and potassium) (Basso et al., 2011).

In the ethanol production process, as soon as the must is added to the yeast, they enter a stage of adaptation and begin to multiply. Subsequently, ethanol is produced and CO is released$_2$. Once the sugar present in the must has been used up, the yeasts begin to reduce their fermentation activity (Paschoalini & Alcarde, 2009).

In Brazilian distilleries, alcoholic fermentation processes are called: continuous fermentations and batch fermentations (with and without centrifugation). There are several types of continuous fermentation processes. It is characterized by having a continuous feed at a constant flow rate, with the volume of the reaction being kept constant through the continuous withdrawal of fermented broth. It is an uninterrupted process that can operate for long periods. The must is mixed with the yeast in the first dome and passes on to the others in a continuous process until it reaches the last dome, where the sugar concentration will be as low as possible (Amorim et al., 1989; Teixeira et al., 2007).

The oldest ethanol production process in Brazil is batch fermentation without yeast recirculation, which is used to produce brandies. In this process, several vats are used, where several fermentations take place. The vats are filled, fermented and processed one by one. Then, after fermentation is complete, the yeast is expected to decant. However, this process can be suitable for the production of brandy and disadvantageous for the production of ethanol, since in order for decanting to take place, bacterial contamination or the presence of contaminating yeasts must occur in order for the yeast (yeast biomass) to flocculate. Due to the lower yields obtained in the batch process, advances were made. This progress was made in the 1930s with the emergence of the Melle-Boinot process (Batch Fermentation with Centrifugation), in which the following were added to the batch process: treatment of the yeast with water and acid, centrifugation and reinsertion of the yeast into the process (cell recycle) (Amorim et al., 1996; Amorim et al., 1989). Since then, this has been the most widely used process in Brazilian distilleries, as it allows the yeast to be used in several fermentations, resulting in a higher yield.

Theoretically, in different fermentation processes, for every 100 kg of ART (Total Reducing Sugars = sucrose, glucose and fructose), a maximum of 64.75 liters of absolute alcohol are produced at a temperature of 20°C (Ghose et al., 1979; Okolo et al., 1987).

According to Basso et al. (2008), cell recycling can cause a drop in cell viability over the course of several fermentation cycles. On the other hand, it can contribute positively to yeast cells, leading to the evolution of multi-tolerant strains, because during cell recycling there is greater selective pressure, leading to greater tolerance of the cells to the stressful conditions of industrial fermentation.

In industrial alcoholic fermentations, a variety of stress factors impose restrictions on yeast cells, interfering with their metabolism and growth, of which we can list: high sugar concentration, presence of salts in the must, high temperature, high ethanol concentration, low external pH (due to acid treatment), bacterial contamination, yeast recycling, nutrient deficiency and presence of inhibitory agents (sulphite) (Graves et al., 2007).

Temperature can directly affect yeast metabolism, growth and cell viability. Yeasts in distilleries grow well at temperatures of 30-33°C. Fermentation can occur more quickly at even higher temperatures, but as temperatures increase, the cells become more sensitive to the toxicity exerted by the excreted ethanol, causing a drop in cell viability (Amorim et al., 1989; Piper, 1995; Aldiguier et al., 2004).

Ethanol stress inhibits the growth and decreases the cell viability of yeasts. High ethanol content can adversely affect the transport system, cell signaling and the activity of key enzymes in the glycolytic pathway. The toxicity of ethanol acts directly on the cytoplasmic membrane of cells, altering the degree of polarity and increasing its fluidity, causing an interruption in cell growth when in high concentrations (Ingran, 1976; Banat et al., 1998).

In industrial fermentations, pH is an important factor in controlling bacterial contamination. The pH values of musts in distilleries range from 4.5 to 5.5, with a good buffering capacity. The intracellular pH remains in the 5.8 to 6.9 range, as long as the extracellular pH is in the 2 to 7 range. However, a lower extracellular pH causes yeasts to spend more energy maintaining the internal pH, as well as affecting the transport proteins of the cytoplasmic membrane (Steckelberg, 2001).

Bacterial contamination is one of the factors that has the greatest deleterious effect on yeast. When there is an increased presence of bacteria, there is a significant formation of lactic and acetic acids produced by the bacteria. The development of bacteria proceeds from a detour of the carbon source, consequently reducing the formation of ethanol (Basso et al., 2008).

Conditions of osmotic stress, such as those found in fermentations with a high alcohol content, can impair yeast metabolism. Increased osmotic pressure in the must can cause a concomitant decrease in viability, growth and fermentation performance. Hyperosmotic shock in *S. cerevisiae* can result in a loss of turgidity due to cytoplasmic water loss. Under conditions of osmotic stress, osmoregulatory mechanisms occur, and these are related to the cellular synthesis of glycerol and trehalose. In this way, cells can prevent dehydration and protect their structures (Mager & Siderius, 2002).

The reserve carbohydrates trehalose and glycogen can account for up to 30% of yeast dry matter. Trehalose is a disaccharide made up of two glucose molecules and is responsible for maintaining viable vegetative cells and spores. The protective function of trehalose is the result of the relationship between its location and cellular mobilization. It accumulates in the cytoplasm, influencing water activity, contributing to cell latency and increasing resistance to water stress (Parrou et al., 1997; Singer & Lindquist, 1998; Zhao & Bai, 2009). Trehalose maintains membrane integrity because it replaces water molecules on both sides of the phospholipid layer in conditions where there is low water activity. Both reserve carbohydrates, trehalose and glycogen, fluctuate during fermentation and can end the process with different levels. At the end of fermentation, these reserves can be metabolized by the cells, producing more ethanol and enriching the yeast with nitrogen (Amorim et al., 1996).

In 1989, the karyotyping technique enabled the rescue of dominant and persistent strains in the industrial process, belonging to the *S. cerevisiae* species and called CAT-1, PE-2, BG-1, SA-1. These strains are currently widely used in Brazilian distilleries and were selected for good fermentation characteristics, such as: high ethanol yield, high cell viability during recycling, high trehalose content, non-foaming and non-flocculating. As well as having desirable fermentation characteristics, they also stood out for their good implantation capacity (60-70%) in industrial processes, with a high dominance rate, reaching the end of the harvest representing almost all of the biomass in the vat (Basso et al., 2008).

2.3.1 Fermentation with high alcohol content

One way of increasing the levels of ethanol obtained in industrial fermentation is by intensifying the process. The technology called *Very High Gravity Fermentation* (*VHG Fermentation*) is an improvement process designed to obtain high ethanol concentrations of up to 15% (v/v) from high concentrations of total sugars (>250 g.L).$^{-1}$

VHG technology allows for the reduction of process water, thus reducing the operating cost of distillation and the generation of vinasse, resulting in significant energy savings (Thomas et al., 1996). However, throughout *VHG* fermentation, industrial yeast encounters significant stress induced by osmotic pressure, which leads to variations in fermentation kinetics. Strains tolerant to osmotic and ethanolic stress are prerequisites for more efficient ethanol production under *VHG fermentation* conditions. The level of ethanol at the end of fermentation depends mainly on the nutritional conditions for cell maintenance. Another common source of variations in kinetics may be related to the quality of the raw material, as the quality of molasses and sugarcane juice can change depending on the species, harvest period, climatic conditions and extraction procedure (Wang et al., 2013).

According to Puligundla et al. (2011), savings of around 40% in water consumption can be achieved by fermentation with a high alcohol content. By using a mash with a high concentration of sugars, considerable energy savings can be made because there is less fluid in the heating, cooling and distillation stages.

2.4 Msn2 and Msn4 transcription factors in *S. cerevisiae*

All structural and functional units of living organisms, i.e. all cells, have the ability to respond to sudden changes in environmental conditions that are capable of threatening cell viability (Kültz, 2005).

Cellular response mechanisms include environmental sensors and signal transduction pathways that can cause significant changes in gene expression programs. The induction or repression of gene expression under stress conditions allows rapid adaptation to various conditions, resulting in increased cellular fitness and survival (Gasch et al., 2000).

Transcription factors promote the expression of different genes in response to various stress conditions. While some promote the transcription of specific sets of genes, thus allowing cellular adaptation to specific stresses, others promote the transcription of many genes in response to a wide variety of environmental stresses (Estruch, 2000). In more detail, transcription factors are proteins that bind to specific sequences (promoters) in the DNA of eukaryotic cells to allow the RNA-polymerase enzyme to bind to the DNA, thus enabling transcription and subsequent translation (Phillips, 2008).

In *S. cerevisiae* yeasts, the transcription factors Hsf1 and Msn2/Msn4 are the main players in the signal transduction systems and play important roles in the general response to stress (Estruch, 2000; Sadeh et al., 2011).

When *S. cerevisiae* cells are faced with stressful conditions, transcription factors are phosphorylated and transported to the nucleus. In this way, they control the regulation of the synthesis of stress genes, because they trigger the transcription of genes with regulatory sequences in their promoters (Estruch, 2000).

For example, in *S. cerevisiae,* the transcription factor Hsf1 (*heat shock factor protein 1*) is related to the heat shock response, since, in response to high temperatures, this transcription factor is phosphorylated and transported to the cell nucleus, triggering the transcription of genes whose promoter sequence contains the heat shock element (Mager & De-Kruijff, 1995). The *MSN2* gene and its homologue *MSN4* also participate in the heat shock response and are normally activated by various other types of stress (Martinez-Pastor et al., 1996). The *MSN2* and *MSN4* genes are similar in that their sequences are 41% identical. The Msn2 transcription factor is dominant and contains two zinc finger motifs (Cys2His2) near the C-terminal tail (Estruch, 2000; Schmitt & Mcentee, 1996; Martinez-Pastor et al., 1996).

The Msn2 and Msn4 transcription factors activate genes containing STRE (*stress-responsive element*) (Smith et al., 1998). The Stress Response Element (STRE) is a functional sequence with 5 base pairs (bp), capable of activating the expression of genes in response to various stimuli harmful to cells (Nicholls et al., 2004). According to Toone & Jones (1998), several genes involved in the general response to stress have been described as containing stress response elements (STREs) in their promoter regions. *TPS1/2* genes, which encode enzymes for trehalose synthesis, or the *GPD1* gene, which encodes an enzyme necessary for glycerol synthesis, are examples of genes regulated by the STRE.

It is important to note that the Msn2 and Msn4 factors are directly related to the levels of cyclic AMP (cAMP) and protein kinase A (PKA), since, under stressful conditions or nutritional deficiency, the intracellular levels of cyclic AMP in *S. cerevisiae* decrease, consequently causing a low activity of protein kinase A (PKA) (Gorner et al., 1998). In this situation, the transcription factors Msn2 and Msn4 are phosphorylated and transported to the nucleus in accordance with the low activity of this protein (PKA) (Martinez-Pastor et al., 1996).

The plasma membrane of *S. cevevisiae* cells contains receptor proteins that signal the availability of nutrients in the extracellular environment. This cellular signaling can regulate the transcription of genes and the translation of proteins (Gancedo, 2008). By interacting with carbohydrates, these membrane receptor proteins cause changes in their structural conformations and trigger an intracellular signaling cascade, promoting the activation of one or more proteins in the intracellular environment (Forsberg & Ljungdahl, 2001).

An example of plasma membrane receptor proteins is the protein called Gprl (*G-protein-coupled receptor*). In the yeast *S. cerevisiae*, the rapid activation of the cyclic AMP pathway by glucose and sucrose requires the presence of Gpr1 (Kraakman et al., 1999; Lorenz et al., 2000), and the affinity of this receptor protein is greater for sucrose (Lemaire et al., 2004). Signaling by Gpr1 occurs through the interaction of this receptor protein with the carbohydrate, thus causing a structural change in the conformation of the membrane receptor proteins. From there, the Gpr1 protein bound to the Gpa2 protein interacts with an enzyme called adenylate cyclase (encoded by the *CYR1* gene), forming cyclic AMP (Lorenz et al., 2000; Colombo et al., 2004; Peeters et al., 2006; Tamaki, 2007). When cyclic AMP is present inside the cell, it activates protein kinases (dependent on cyclic AMP). These proteins are responsible for phosphorylating a series of other proteins, activating or inhibiting them (Sreenath et al., 1988). Protein kinase A (PKA) is activated by an increase in intracellular cyclic AMP and phosphorylates proteins that can repress or induce the expression of various genes (Thevelein & Winde, 1999).

According to Reinders et al. (1998) and Tamaki (2007), the activation of protein kinase A (PKA) by cyclic AMP in *S. cerevisiae* yeast is necessary for the proper regulation of growth, cell cycle progression and development in response to nutritional conditions. In addition, high PKA activity also causes repression of the *TPS1* and *TPS2* genes that encode the enzyme trehalose synthase (Winderickx et al., 1996; Trevisol et al., 2014).

According to Thevelein (1994) and Rolland et al. (2000), there are two essential requirements for the activation of cyclic AMP induced by glucose/sucrose: the first requirement is the availability of glucose/sucrose in the extracellular environment, signaled by the receptor protein Gpr1, and the second is the availability of intracellular glucose, requiring the enzymes hexokinases, since according to the authors there is a requirement for phosphorylation of glucose (first stage of glycolysis), promoting the activation of cyclic AMP.

The addition of glucose or sucrose to yeast cells grown on a non-fermentable carbon source or in the stationary phase causes a transient surge in intracellular cyclic AMP levels (Mbonyi et al., 1990; Kraakman et al., 1999). Other sugars, such as fructose, maltose, maltotriose and galactose, cannot trigger a strong cyclic AMP and PKA response (Rolland et al., 2000; Rolland et al., 2001). Consequently, during the growth of *S. cerevisiae* in a rich medium containing glucose/sucrose, the expression of genes under the control of the Msn2 and Msn4 transcription factors is repressed, with expression being induced only in the transition phase to respiratory metabolism, in conditions where there is nitrogen limitation, and also by various other types of stress. The Msn2 and Msn4 factors can modulate the expression of more than 200 genes in response to harmful stimuli, such as high ethanol content, oxidative stress, lack of nutrients, low pH, high temperatures, osmotic pressure, among others (Martinez-Pastor et al., 1996).

According to Trott & Morano (2003), the *MSN2* and *MSN4* genes can be regulated according to nutritional status and cell growth phase. The stationary and decline phases can cause these genes to be induced.

The figure below (Figure 1) shows the targets of the cAMP and PKA nutrient signaling pathways. The targets include key proteins involved in cell growth control, glucose metabolism, resistance to some types of stress (including trehalose metabolism), flocculation, filamentous growth and synthesis of volatile esters (Verstrepen et al., 2004).

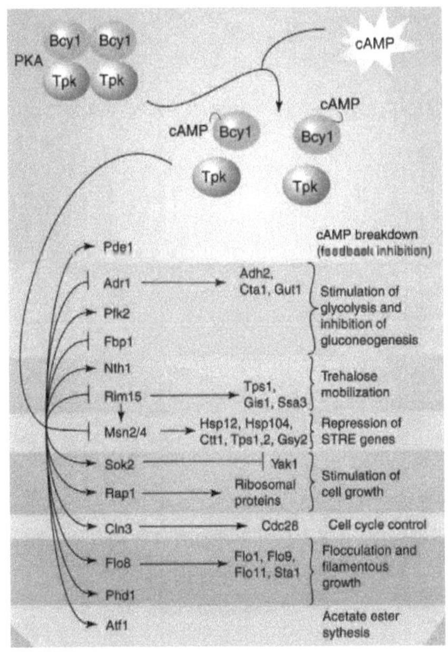

Figure 1 - Targets of the cAMP and PKA nutrient signaling pathways (Verstrepen et al., 2004)

2.5 Tryptophan biosynthesis and amino acid supplementation in alcoholic fermentation

Studies have shown that there is a relationship between amino acid biosynthesis and tolerance to ethanolic stress in yeasts (Takagi et al., 2005; Hirasawa et al., 2007; Li et al., 2010). According to Hirasawa et al. (2007), overexpression of the *TRP1* and *TRP5* genes, involved in tryptophan biosynthesis, led to increased tolerance to ethanol stress (5% v/v) in *S. cerevisiae* yeasts grown in the laboratory. Tolerance to high concentrations of ethanol is one of the most desirable characteristics for the industry, so the selection of ethanol-tolerant mutants could have direct application in the fermentation process (Yazawa et al., 2007). Through DNA *microarray* analysis, it was reported that increased expression of genes related to tryptophan biosynthesis confers greater tolerance to ethanol stress for yeast cells, proving the fact that laboratory strains with over-expression of tryptophan biosynthesis genes could show greater tolerance to ethanol (Hirasawa et al. 2007). In the work by Yoshikawa et al. (2009), the growth behavior of yeast strains with a single deletion in the *TRP1* gene under ethanol stress was verified, and it was found that the strains with deletions in the genes classified in the "tryptophan metabolism" category were sensitive to 8% (v/v) ethanol. It is therefore possible that tryptophan synthesis genes may contribute to the yeast cell's greater tolerance to ethanol stress.

Amino acid supplementation has also been studied for significantly improving the tolerance and biomass formation of *S. cerevisiae*, especially under stress conditions such as those caused by high levels of ethanol (Thomas & Ingledew, 1990; Thomas & Ingledew, 1992, Chen et al., 1993). According to Pham & Wright (2008), the supplementation of amino acids in fermentation can lead to positive cellular responses, including a reduction in the Lag phase and greater cell viability in media with a high sugar concentration. In the study by Hu et al. (2005), the amino acids isoleucine, methionine and phenylalanine increased the tolerance of *S. cerevisiae* under conditions of ethanol stress, and the greater tolerance of the cells was due to the incorporation of the supplementary amino acids into the cytoplasmic membrane, increasing the capacity to reduce the fluidizing effect exerted by ethanol. Some authors have reported that glutamine, asparagine and ammonium are the preferred nitrogen sources for the yeast *S. cerevisiae*. These nitrogenous compounds are used primarily, supporting higher growth rates than other nitrogen sources, to which the yeast has a secondary preference (Ter- Schure et al., 2000; Dubois & Messenguy, 1997; Wiame et al., 1985; Cooper, 1982). According to Basso & Amorim (2001), the yeast *S. cerevisiae* uses nitrogen preferentially in the ammoniacal form (NH_4^+) to synthesize amino acids and nitrogenous bases necessary for its growth. In conditions of ammonium ion deficiency, the yeast uses other pathways to metabolize amide or amino N.

In yeast, amino acids can be incorporated directly into the biomass during consumption (Albers et al., 1996), but they can also help in the production of proteins and the formation of structural components. Amino acids are therefore referred to as the "building blocks" of proteins. In addition, they play a central role in the general metabolism of yeasts. Studies related to determining the metabolic pathways of amino acid utilization are considered to be highly complex. This is because they involve describing the pathways by which amino acids are used as carbon and nitrogen sources, their biosynthesis, as well as their conversion into other metabolites, including nucleotides (Ljungdahl & Daignan-Fornier, 2012).

In cells, sensors located in the cytoplasmic membrane respond to the availability of different sets of nutrients, including different sources of nitrogen. These environmental sensors operate in conjunction with networks of intracellular sensing systems. In addition, catabolic and anabolic pathways generate multiple metabolic intermediates that contribute significantly to the complexity of the chemical composition of cells (Zaman et al., 2008).

The presence of external amino acids induces the expression of several permeases of broad specificity. This transcription response is mediated by the Ssy1-Ptr3-Ssy5 (SPS) sensor located on the plasma membrane (Ljungdahl, 2009). Once internalized, amino acids can be used in biosynthetic processes, be deaminated to generate ammonium, or serve as substrates for transaminases that transfer amino groups to a-ketoglutarate to form glutamate (Cooper, 1982; Magasanik & Kaiser, 2002). In yeasts grown on glucose media, ammonium can be assimilated by two anabolic reactions. In the first reaction, glutamate is synthesized from ammonium and catalyzed a-ketoglutarate (by NADPH-dependent glutamate dehydrogenase), and in the second, glutamine is synthesized from ammonium and glutamate (by glutamine synthetase) (Avendano et al., 1997; De-Luna et al., 2001).

According to Walker (1998), nitrogen makes up approximately 10% of the dry weight of yeast cells. To supplement nitrogen, ammonium sulphate is commonly used by the industry. Yeast grows best in the presence of amino acids because they are the constituents of proteins. If the substrate supplied is deficient in amino acids, a series of fermentation problems occur. Even when the yeast can produce its own amino acids, it is generally better for the cell to assimilate them.

References

Albers, E.; Larsson, C.; Lidén, G.; Niklasson, C.; Gustafsson, L. Influence of nitrogen source on *Saccharomyces cerevisiae* anaerobic growth and product formation. **Applied and Environmental Microbiology**, v. 62, p. 3187-3195, 1996.

Alexandre, H.; Ansanay-Galeote, V.; Dequin, S.; Blondin, B. Global gene expression during short-term ethanol stress in *Saccharomyces cerevisiae*. **FEBS Letters**, v. 498, n. 1, p. 98-103, 2001.

Aldiguier, A.S.; Alfenore, S.; Cameleyre, X.; Goma, G.; Uribelarrea, J.L.; Guillouet, S.E.; Molina-Jouve, C. Synergistic temperature and ethanol effect on *Saccharomyces cerevisiae* dynamic behavior in ethanol bio-fuel production. **Bioprocess and Biosystems Engineering**, v. 26, n. 4, p. 217222, 2004.

Alfenore, S.; Molina-Jouve, C.; Guillouet, S; Uribelarrea, J.L.; Goma, G.; Benbadis, L. Improving ethanol production and viability of *Saccharomyces cerevisiae* by a vitamin feeding strategy during fed-batch process. **Applied Microbiology and**

Biotechnology, v. 60, n. 1-2, p. 6772, 2002.

Amorim, H.V.; Basso, L.C.; Alves, D.M.G. **Alcohol production processes, control and monitoring.** Piracicaba: FERMENTEC/FEALQ/ESALQ-USP, 1996. 103p.

Amorim, H.V.; Oliveira, A.J.; Zago, E.A.; Basso, L.C.; Gallo, C.R. **Processes of alcoholic fermentation, their control and monitoring**. Piracicaba: FERMENTEC/CEBTEC/ESALQ-USP, 1989. 145p.

Andrietta, M.G.S.; Andrietta, S.R.; Steckelberg, C.; Stupielo, E.N.A. Bioethanol - 30 years of proàlcool. **International Sugar Journal**, v. 109, n. 1299, p. 195-200, 2007.

Avendano, A.; Deluna, A.; Olivera, H.; Valenzuela, L.; Gonzalez, A. GDH3 encodes a glutamate dehydrogenase isozyme, a previously unrecognized route for glutamate biosynthesis in *Saccharomyces cerevisiae*. **Journal of Bacteriology**, v. 179, p. 5594-5597, 1997.

Bai, F.W.; Anderson, W.A.; Moo-Young, M. Ethanol fermentation technologies from sugar and starch feedstocks. **Biotechnology Advances**, v. 26, n. 1, p. 89-105, 2008.

Bakker, B.M.; Overkamp, K.M.; Van-Maris, A.J.A.; Kotter, P.; Luttik, M.A.H.; Van-Dijken, J.P.; Pronk, J.T. Stoichiometry and compartmentation of NADH metabolism in *Saccharomyces cerevisiae*. **FEMS Microbiology Reviews**, v. 25, n. 1, p. 15-37, 2001.

Balat, M.; Balat, H. Recent trends in global production and utilization of bio-ethanol fuel. **Applied Energy**, London, v. 86, p. 22732282, 2009.

Banat, I.M.; Nigam, P.; Singh, D.; Marchant, R.; Mchale, A.P. Review: Ethanol production at elevated temperatures and alcohol concentrations: Part I - Yeasts in general. **World Journal of Microbiology & Biotechnology**, v. 14, n. 6, p. 809-821, 1998.

Barnett, J. A. The utilization of sugars by yeasts. **Advances in Carbohydrate Chemistry and Biochemistry**, v. 32, p. 125-234, 1976.

Barnett, J.A. The utilization of disaccharides and some other sugars by yeasts. **Advances in Carbohydrate Chemistry and Biochemistry**, v. 39, p. 347-404, 1981.

Basso, L.C. Amorim, H.V.; Oliveira, A.J.; Lopes, M.L. Yeast selection for fuel ethanol in Brazil. **FEMS Yeast Research**, Amsterdam, v. 8, n. 7, p. 1155-1163, 2008.

Basso, L.C.; Basso, T.O.; Rocha, S.N. Ethanol production in Brazil: the industrial process and its impact on yeast fermentation. In: SANTOS BERNARDES, M.A. dos (Ed.). **Biofuel Production - Recent Developments and Prospects**. Croatia: INTECH, 2011a. chap. 5, p. 85100.

Blomberg, A.; Adler, L. Physiology of osmotolerance in fungi. **Advances in Microbial Physiology**, v. 22, p. 145-212, 1992.

Bücker, A. **Genomic engineering of an industrial strain of *Saccharomyces cerevisiae* to improve ethanol tolerance**. 2014. 111p. Thesis (PhD in Biochemistry) - Federal University of Santa Catarina, Florianópolis, 2014.

Cardona, C.A.; Sanchez, O.J. Fuel ethanol production: Process design trends and integration opportunities. **Bioresource Technology**, v. 98, n. 12, p. 2415-24-57, 2007.

Castrillo, J.I.; Ugalde, U.O. A general model of yeast energy metabolism in aerobic chemostat culture. **Yeast**, Chichester, v. 10, p. 185197, 1994.

Ceccato-Antonini, S.R. Biotechnological implications of filamentation in *Saccharomyces cerevisiae*. **Biotechnology Letters**, v. 30, n. 7, p. 1151-1161, 2008.

Cerqueira-Leite, R.C.; Leal, M.R.L.V.; Cortez, L.A.B.; Griffin W.M.; Scandiffio, M.I.G. Can Brazil replace 5% of the 2025 gasoline world demand with ethanol? **Energy**, v. 34, n. 5, p. 655-661, 2009.

Chen, J.C.P.; Chou, C.C. Cane Sugar Handbook - A Manual for Cane Sugar Manufacturers and Their Chemists. In: Chen, J.C.P.; Chou, C.C. (Eds.). **Sugars and non-sugars in sugarcane**. John Wiley & Sons Inc, 1993, chap. 2, p. 21-30.

Chen, Y.; Kirk, N.; Piper, P.W. Effects of medium composition on MFα1 promoter-directed secretion of a small protease inhibitor in *Saccharomyces cerevisiae* batch fermentation. **Biotechnology Letters**, v. 15, p. 223-228, 1993.

Colombo, S.; Ronchetti, D.; Thevelein, J.M.; Winderickx, J.; Martegani, E. Activation state of the Ras2 protein and glucose-induced signaling in *Saccharomyces cerevisiae*. **The Journal of Biological Chemistry**, v. 279, n. 45, p. 46715-46722, 2004.

Cooper, T.G. Nitrogen metabolism in *Saccharomyces cerevisiae*. In: Strathern, J.N.; Jones, E.W.; Broach, J.R. (Eds.). **The molecular and cellular biology of the yeast *Saccharomyces***. Cold Spring Harbor Laboratory, 1982, v. 11B, chap. 2, p. 39-99.

Cronwright, G.R.; Rohwer, J.M.; Prior, B.A. Metabolic control analysis of glycerol synthesis in *Saccharomyces cerevisiae*. **Applied and Environmental Microbiology**, v. 68, n. 9, p. 4448-4456, 2002.

DeLuna, A.; Avendano, A.; Riego, L.; Gonzalez, A. NADP- glutamate dehydrogenase isoenzymes of *Saccharomyces cerevisiae*. Purification, kinetic properties, and physiological roles. **The Journal of Biological Chemistry**, v. 276, p. 43775-43783, 2001.

Demirbas, M.F.; Balat, M. Recent advances on the production and utilization trends of bio-fuels: A global perspective. **Energy Conversion and Management,** Oxford, v. 47, p. 2371-2381, 2006.

Dequin, S. The potential of genetic engineering for improving brewing, wine-making and baking yeasts. **Applied Microbiology and Biotechnology**, v. 56, n. 5-6, p. 577-588, 2001.

Duarte, C.G.; Gaudreau, K.; Gibson, R.B.; Malheiros, T.F. Sustainability assessment of sugarcane-ethanol production in Brazil: A case study of a sugarcane mill in Sao Paulo state. **Ecological Indicators**, v. 30, p. 119-129, 2013.

Dubois, E.; Messenguy, F. Integration of the multiple controls regulating the expression of the arginase gene CAR1 of *Saccharomyces cerevisiae* in response to different nitrogen signals: role of Gln3p, ArgRp- Mcm1p, and Ume6p. **Molecular Genetics and Genomics**, v. 253, p. 568580, 1997.

Estruch, F. Stress-controlled transcription factors, stress-induced genes, and stress tolerance in budding yeast. **FEMS Microbiology Reviews**, v. 24, p. 469-486, 2000.

Estruch, F. Stress-controlled transcription factors, stress-induced genes and stress tolerance in budding yeast. **FEMS Microbiology Reviews**, v. 24, p. 469-486, 2000.

Van-Voorst, F.; Houghton-Larsen, J.; J0nson, L.; Kielland-Brandt, M.C.; Brandt, A. Genome-wide identification of genes required for growth of *Saccharomyces cerevisiae* under ethanol stress. **Yeast**, v. 23, n.5, p. 351359, 2006.

Forsberg, H.; Ljungdahl, P.O. Sensors of extracellular nutrients in *Saccharomyces cerevisiae*. **Current Genetics**, v. 40, n. 2, p. 91-109, 2001.

Gallo, M. A fuel surcharge policy for reducing road traffic greenhouse gas emissions. **Transport Policy**, v. 18, n. 2, p. 413-424, 2011.

Gancedo, J.M. The early steps of glucose signalling in yeast. **FEMS Microbiology Reviews**, v. 32, n. 4, p. 673-704, 2008.

Gasch, A.P.; Spellman, P.T.; Kao, C.M.; Carmel-Harel, O.; Eisen, M.B.; Storz, G.; Botstein, D.; Brown, P.O. Genomic expression programs in the response of yeast cells to environmental changes. **Molecular Biology of the Cell**, v. 11, n. 12, p. 4241-4257, 2000.

Gascón, S.; Lampen, J.O. Purification of the Internal Invertase of Yeast. **The Journal of Biological Chemistry**, v. 243, n. 7, p. 1567-1572, 1968.

Gerbens-Leenes, P.W.; Van Lienden; Hoekstra, A.Y.; Van Der Meer, T.H. Biofuel scenarios in a water perspective: The global blue and green water footprint of road transport in 2030. **Global Environmental Change**, v. 22, n. 3, p. 764-775, 2012.

Ghose, T.K.; Tyagi, R.D. Rapid ethanol fermentation of cellulose hydrolysate: batch versus continuous systems. **Biotechnology and Bioengineering**, v. 21, p. 1387-1400, 1979.

Goldemberg, J. Biomass and energy. **Quimica Nova**, Sao Paulo, v. 32, n. 3, p. 582-587, 2009.

Goldemberg, J. Ethanol for a Sustainable Energy Future. **Science**, v. 315, n. 5813, p. 808-810, 2007.

Goldemberg, J.; Coelho, S.T.; Lucon, O. How adequate policies can push renewables. **Energy Policy**, v. 32, n. 9, p. 1141-1146, 2004.

Golubev, W.I. Antagonistic interactions among yeasts. In: ROSA, C.A.; PETER, G. (Eds.). **The Yeast Handbook: Biodiversity and Ecophysiology of yeasts**. Heidelberg: Springer, 2006, chap. 10, p. 197-219

Gonçalves, R.C.; Siqueira, F.L.T.; Mata, J.F.; Vieira, G.E.G. Challenges and prospects for ethanol production in Brazil - a review. **Revista Liberato**, v. 12, n. 18, p. 107-206, 2011.

Gorner, W.; Durchschlag, E.; Martinez-Pastor, M.T.; Estruch, F.; Ammerer, G.; Hamilton, B.; Ruis, H.; Schüller, C. Nuclear localization of the C2H2 zinc finger protein Msn2p is regulated by stress and protein kinase A activity. **Genes & Development**, v. 12, n. 4, p. 586-597, 1998.

Graves, T.; Narendranath, N.; Power, R. Development of a "Stress Model" Fermentation System for Fuel Ethanol Yeast Strains. **Journal of the Institute of Brewing**, v. 113, n. 3, p. 263-271, 2007.

Gray, K.A.; Zhao, L.; Emptage, M. Bioethanol. **Current Opinion in Chemical Biology**, v. 10, n. 2, p. 141-146, 2006.

Guardabassi, P.; Goldemberg, J. The Prospects of first generation ethanol in developing countries. **Plants and BioEnergy**, v. 4, p 3-11, 2014.

Guo, Z.P.; Zhang, L.; Ding, Z.Y.; Shi, G.Y. Minimization of glycerol synthesis in industrial ethanol yeast without influencing its fermentation performance. **Metabolic Engineering**, v. 13, n. 1, p. 49-59, 2011.

Hirasawa, T; Yoshikawa, K.; Nakakura, Y.; Nagahisa, K.; Furusawa, C.; Katakura,

Y.; Shimizu, H.; Shioya, S. Identification of target genes conferring ethanol stress tolerance to *Saccharomyces cerevisiae* based on DNA microarray data analysis. **Journal of Biotechnology**, v. 131, n. 1, p. 34-44, 2007.

Heux, S.; Cachon, R.; Dequin, S. Cofactor engineering in *Saccharomyces cerevisiae*: Expression of a H2O-forming NADH oxidase and impact on redox metabolism. **Metabolic Engineering**, v. 8, n. 4, p. 303-314, 2006.

Hu, C.K.; Bai, F.W.; An, L.J. Protein amino acid composition of plasma membranes affects membrane fluidity and thereby ethanol tolerance in a self-flocculating fusant of Schizosaccharomyces pombe and

Saccharomyces cerevisiae. **Sheng Wu Gong Cheng Xue Bao**, v. 21, n. 5, p. 809-813, 2005.

Ingram, L.O. Adaptation of membrane lipids to alcohols. **Journal of Bacteriology**, v. 125, n. 2, p. 670-678, 1976.

Iwahashi, H.; Obuchi, K.; Fujii, S.; Komatsu, Y. The correlative evidence suggesting that trehalose stabilizes membrane structure in the yeast *Saccharomyces cerevisiae*. **Cellular and Molecular Biology**, v. 41, n. 6, p. 763-769, 1995.

Kraakman, L.; Lemaire, K.; Ma, P.; Teunissen, A.W.; Donaton, M.C.; Van-Dijck, P.; Winderickx, J, De-Winde, J.H.; Thevelein, J.M. A *Saccharomyces cerevisiae* G-protein coupled receptor, Gpr1, is specifically required for glucose activation of the cAMP pathway during the transition to growth on glucose. **Molecular Microbiology**, v. 32, n. 5, p. 1002-1012, 1999.

Kültz, D. Molecular and evolutionary basis of the cellular stress response. **Annual Review of Physiology**, v. 67, p. 225-257, 2005.

Lagunas, R. Energy metabolism of *Saccharomyces cerevisiae* discrepancy between ATP balance and known metabolic functions. **Biochimica et Biophysica Acta (BBA) - Bioenergetics**, v. 440, n. 3, p. 661-674, 1976.

Lagunas, R. Is *Saccharomyces cerevisiae* a typical facultative anaerobe? **Trends in Biochemical Sciences**, v. 6, p. 201-203, 1981.

Lagunas, R. Sugar transport in *Saccharomyces cerevisiae*. **FEMS Microbiology Letters**, v. 104, n. 3-4, p. 229-242, 1993.

Lemaire, K.; Van-De-Velde, S.; Van-Dijck, P.; Thevelein, J.M. Glucose and sucrose Act as agonist and mannose as antagonist ligands of the G protein-coupled receptor Gpr1 in the yeast *saccharomyces cerevisiae*. **Molecular Cell**, v. 16, n. 2, p. 293-299, 2004.

Lewis, J. G. Learmonth, R. P.; Watson, K. Induction of heat, freezing and salt tolerance by heat and salt shock in *Saccharomyces cerevisiae*. **Microbiology**, v. 141, p. 687-694, 1995.

Li, B.Z.; Cheng, J.S.; Ding, M.Z.; Yuan, Y.J. Transcriptome analysis of differential responses of diploid and haploid yeast to ethanol stress. **Journal of Biotechnology**, v. 148, n. 4, p. 194-203, 2010.

Ljungdahl, P.O. Amino-acid-induced signalling via the SPS-sensing pathway in yeast. **Biochemical Society Transactions**, v. 37, p. 242-247, 2009.

Ljungdahl, P.O.; Daignan-Fornier, B. Regulation of amino acid, nucleotide, and phosphate metabolism in *Saccharomyces cerevisiae*. **Genetics**, v. 190, n. 3, p. 885-929, 2012.

Lorenz, M.C.; Pan, X.; Harashima, T.; Cardenas, M.E.; Xue, Y.; Hirsch, J.P.; Heitman, J. The G protein-coupled receptor gpr1 is a nutrient sensor that regulates pseudohyphal differentiation in *Saccharomyces cerevisiae*. **Genetics**, v. 154, n. 2, p. 609-622, 2000.

Macedo, I.C. Current situation and prospects for ethanol. **Estudos Avançados**, v. 21, n. 59, p. 157-165, 2007.

Madigan, M.T.; Martinko, J.M.; Stahl, D.; Clark, D.P. Eukaryotic cell biology and eukaryotic microorganisms. In: Madigan, M.T.; Martinko, J.M.; Stahl, D.; Clark, D.P. (Eds.). **Brock Biology of Microorganisms**. Benjamin Cummings, 2000, chap. 20, p. 584-612.

Magasanik, B.; Kaiser, C.A. Nitrogen regulation in *Saccharomyces cerevisiae*. **Gene**,

v. 290, n. 1-2, p. 1-18, 2002.

Mager, W.H.; Siderius, M. Novel insights into the osmotic stress response of yeast. **FEMS Yeast Research**, v. 2, n. 3, p. 251-257, 2002.

Mager, W.; De-Kruijff, A.J.J. Stress-induced transcriptional activation. **Microbiological Reviews**, v. 59, n. 3, p. 506-531, 1995.

Mayerhoff, Z. D. V. L. Bioethanol patents highlight Brazilian development. **Inovaçâo Uniemp**, Campinas, v. 2, n. 2, p. 2223, 2006.

Martinez-Pastor, M.T.; Marchler, G.; Schüller, C.; Marchler-Bauer, A.; Ruis, H.; Estruch, F. The *Saccharomyces cerevisiae* zinc finger proteins Msn2p and Msn4p are required for transcriptional induction through the stress response element (STRE). **The EMBO Journal**, v. 15, n. 9, p. 22272235, 1996.

Martinez-Pastor, M.T.; Marchler, G.; Schüller, C.; Marchler-Bauer, A.; Ruis, H.; Estruch, F. The *Saccharomyces cerevisiae* zinc finger proteins Msn2p and Msn4p are required for transcriptional induction through the stress response element (STRE). **The EMBO Journal**, v. 15, n. 9, p. 22272235, 1996.

Mbonyi, K.; Van-Aelst, L.; Argüelles, J.C.; Jans, A.W.; Thevelein, J.M. Glucose-induced hyperaccumulation of cyclic AMP and defective glucose repression in yeast strains with reduced activity of cyclic AMPdependent protein kinase. **Molecular and Cellular Biology**, v. 10, n. 9, p. 4518-4523, 1990.

Mckendry, P. Energy production from biomass (part 2): conversion technologies. **Bioresource Technology**, v. 83, n. 1, p. 47-54, 2002.

Nicholls, S.; Straffon, M; Enjalbert, B.; Nantel, A.; Macaskill, S.; Whiteway, M.; Brown, A.J.P. Msn2- and Msn4-like transcription factors play no obvious roles in the stress responses of the fungal pathogen *Candida albicans*. **Eukaryot Cell**, v. 3, n. 5, p. 1111-1123, 2004.

Nigam, P.S.; Singh, A. Production of liquid biofuels from renewable resources. **Progress in Energy and Combustion Science**, v. 37, n. 1, p. 52-68, 2011.

Okolo, B.; Johnston, J.R.; Berry, D.R. Toxicity of ethanol, n-butanol and iso-amyl alcohol in *Saccharomyces cerevisiae* when supplied separately and in mixtures. **Biotechnology Letters**, v. 9, n. 6, p. 431-434, 1987.

Ostergaard, S.; Olsson, L.; Nielsen, J. Metabolic Engineering of *Saccharomyces cerevisiae*. **Microbiology and Molecular Biology Reviews**, v. 64, n. 1, p. 34-50, 2000.

Puligundla, P.; Smogrovicova, D.; Obulam, V.S.; Ko, S. Very high gravity (VHG) ethanolic brewing and fermentation: a research update. **Journal of Industrial Microbiology and Biotechnology**, v. 38, n. 9, p. 1133-1144, 2011.

Parrou, J.L.; Teste, M.A.; François, J. Effects of various types of stress on the metabolism of reserve carbohydrates in *Saccharomyces cerevisiae*: genetic evidence for a stress-induced recycling of glycogen and trehalose. **Microbiology**, v. 143, n. 6, p. 1891-1900, 1997.

Park, J.I.; Grant, C.M.; Attfield, P.V.; Dawes, I.W. The freeze-thaw stress response of the yeast *Saccharomyces cerevisiae* is growth phase specific and is controlled by nutritional state via the RAS-cyclic AMP signal transduction pathway. **Applied and Environmental Microbiology**, v. 63, p. 3818-3824, 1997.

Paschoalini, G.; Alcarde, V.E. Study of the fermentation process in a sugar-alcohol plant and a proposal for its optimization. **Revista de Ciência & Tecnologia**, v. 16, n. 32, p. 59-68, 2009.

Peeters, T.; Louwet, W.; Geladé, R.; Nauwelaers, D.; Thevelein, J.M.; Versele, M. Kelch-repeat proteins interacting with the Galpha protein Gpa2 bypass adenylate cyclase for direct regulation of protein kinase A in yeast. **Proceedings of the National Academy of Sciences USA**, v. 103, n. 35, p. 13034-13039, 2006.

Penido-Filho, P. Fuel alcohol. In: NOBEL (Ed.). **Fuel alcohol: obtaining and application in engines**. Sao Paulo: Livraria Nobel S.A., 1980, chap. 3, p. 43-48.

Phillips, T. Transcription factors and transcriptional control in eukaryotic cells. **Nature Education**, v. 1, n. 1, p. 119, 2008.

Piper, P.W. Molecular events associated with acquisition of heat tolerance by the yeast *Saccharomyces cerevisiae*. **FEMS Microbiology Reviews**, v. 11, n. 4, p. 339-356, 1993.

Pronk, J.T.; Yde-Steensma, H.; Van-Dijken, J.P. Pyruvate metabolism in *Saccharomyces cerevisiae*. **YEAST**, v. 12, p. 1607-1633, 1996.

Querol, A.; Fernândez-Espinar, M.T.; Olmo, M.L.D.; Barrio, E. Adaptive evolution of wine yeast. **International Journal of Food Microbiology**, v. 86, n. 1-2, p. 3-10, 2003.

Reed, G.; Nagodawithana, T.W. Yeast-derived products. In: Reed, G.; Nagodawithana, T.W. (Eds.). **Yeast Technology**. New York: Van Nostrand Reinhold, 1990, chap. 8, p. 369-412.

Reinders, A.; Bürckert, N.; Boller, T.; Wiemken, A.; De-Virgilio, C. *Saccharomyces cerevisiae* cAMP-dependent protein kinase controls entry into stationary phase through the Rim15p protein kinase. **Genes & Development**, v. 12, n. 18, p. 2943-2955, 1998.

RFA, Renewable Fuels Association. **World Fuel Ethanol Production.**Available at: <http://www.ethanolrfa.org/resources/industry/statistics> Accessed on: 28 Sept. 2017.

Rocha-Leao, M.H.M.; Panek, A.D.; Costa-Carvalho, V.L.A. Glycogen accumulation during growth of *Saccharomyces cerevisiae*: catabolite repression effects. **IRCS Medical Science-Biochemistry**, v. 12, n. 5, p. 411-412, 1984.

Rolland, F.; De-Winde, J.H.; Lemaire, K.; Boles, E.; Thevelein, J.M.; Winderickx, J. Glucose-induced cAMP signalling in yeast requires both a G-protein coupled receptor system for extracellular glucose detection and a separable hexose kinase-dependent sensing process. **Molecular Microbiology**, v. 38, n. 2, p. 348-358, 2000.

Rolland, F.; Wanke, V.; Cauwenberg, L.; Ma, P.; Boles, E.; Vanoni, M.; De-Winde, J.H.; Thevelein, J.M.; Winderickx, J. The role of hexose transport and phosphorylation in cAMP signalling in the yeast *Saccharomyces cerevisiae*. **FEMS**

Yeast Research**, v. 1, n. 1, p. 33-45, 2001.

Rosa, S.E.S.; Garcia, J.L.F. Second generation ethanol: limits and opportunities. **Revista do BNDES**, v. 32, p. 117-156, 2009.

Sadeh, A.; Movshovich, N.; Volokh, M.; Gheber, L.; Aharoni, A. Fine-tuning of the Msn2/4-mediated yeast stress responses as revealed by systematic deletion of Msn2/4 partners. **Molecular Biology of the Cell**, v. 22, n. 7, p. 3127-3138, 2011.

Schmitt, A.P.; Mcentee, K. Msn2p, a zinc finger DNA-binding protein, is the transcriptional activator of the multistress response in

Saccharomyces cerevisiae. **Proceedings of the National Academy of Sciences**, v. 93, p. 5777-5782, 1996.

Searchinger, T.; Heimlich, R.; Houghton, R.A.; Dong, F.; Elobeid, A.; Fabiosa, J.; Tokgoz, S.; Hayes, D.; Yu, T. Use of U.S. croplands for biofuels increases greenhouse gases through emissions from land-use change. **Science**, v. 319, n. 5867, p. 1238-1240, 2008.

Singer, M.; Lindquist, S. Thermotolerance in *Saccharomyces cerevisiae*: the Yin and Yang of trehalose. **Trends in Biotechnology**, v. 16, n. 11, p. 460-468, 1998.

Smith, A.; Ward, M.P.; Garrett, S. Yeast PKA represses Msn2p/Msn4p-dependent gene expression to regulate growth, stress response and glycogen accumulation. **The EMBO Journal**, v. 17, n. 13, p. 3556-3564, 1998.

Sreenath, T.L.V.; Diego, B.; Nadera, A.; Ravi, D. Two different protein kinase activities phosphorylate RAS2 protein in *saccharomyces cerevisiae*. **Biochemical and Biophysical Research Communications**, v. 157, n. 3, p. 1182-1189, 1988.

Steckelberg, C. **Characterization of yeasts from alcoholic fermentation processes using attributes of cellular composition and kinetic characteristics**. 2001. 202p. Thesis (Doctorate in Chemical Engineering) - Faculty of Chemical Engineering, State University of Campinas, Campinas, 2001.

Takagi, H.; Takaoka, M., Kawaguchi, A.; Kubo, Y. Effect of L- proline on sake brewing and ethanol stress in *Saccharomyces cerevisiae*. **Applied and Environmental Microbiology**, v. 71, p. 8656-8662, 2005.

Tamaki, H. Glucose-stimulated cAMP-protein kinase a pathway in yeast *Saccharomyces cerevisiae*. **Journal of Bioscience and Bioengineering**, v. 104, n. 4, p. 245-250, 2007.

Teixeira, J.A.; Fonseca, M.M.; Vicente, A.A. Geometries and Modes of Operation. In: Fonseca, M.M. (Ed.). **Biological Reactors: Fundamentals and Applications**. Lisbon: LIDEL, 2007, chap. 2, p. 27-68.

Thevelein, J.M. Signal transduction in yeast. **YEAST**, v. 10, n. 13, p. 1753-1790, 1994.

Thevelein, J.M.; Winde, J.H. Novel sensing mechanisms and targets for the cAMP-protein kinase A pathway in the yeast *Saccharomyces cerevisiae*. **Molecular Microbiology**, v. 33, n. 5, p. 904-918, 1999.

Thomas, K.C.; Hynes, S.H.; Ingledew, W.M. Practical and theoretical considerations in the production of high concentration of alcohol by fermentation. **Process Biochemistry**, London, v. 31, n. 4, p. 321-331, 1996.

Thomas, K.C.; Ingledew, W.M. Fuel alcohol production: effects of free amino nitrogen on fermentation of very-high-gravity wheat mashes. **Applied and Environmental Microbiology**, v. 56, p. 2046-2050, 1990.

Thomas, K.C.; Ingledew, W.M. Relationship of low lysine and high arginine concentrations to efficient ethanolic fermentation of wheat mashes. **Canadian Journal of Microbiology**, v. 38, p. 626-634, 1992.

Toone, W.M.; Jones, N. Stress-activated signalling pathways in yeast. **Genes to Cells**, v. 3, p. 485-498, 1998.

Trevelyan, W.E.; Harrison, J.S. Studies on yeast metabolism. 5. The trehalose content of baker's yeast during anaerobic fermentation. **The Biochemical Journal**, London, v. 62, n. 2, p. 177-183, 1956.

Trevisol, E.T.V.; Panek, A.D.; De-Mesquita, J.F.; Eleutherio, E.C.A. Regulation of the yeast trehalose-synthase complex by cyclic AMPdependent phosphorylation. **Biochimica et Biophysica Acta (BBA) - General Subjects**, v. 1840, n. 6, p. 1646-1650, 2014.

Trott, A.; Morano, K.A. The yeast response to heat shock. In: HOHMANN, S.; MAGER, W.H. (Eds.). **Yeast Stress Responses**. Heidelberg: Springer, 2003, chap. 3, p. 71-119.

Van-Dijken, J.P.; Scheffers, W.A. Redox balances in the metabolism of sugars by yeasts. **FEMS Microbiology Letters**, v. 32, n. 3-4, 1986.

Verstrepen, K.J.; Iserentant, D.; Malcorps, P.; Derdelinckx, G.; Van- Dijck, P. Winderickx, J.; Pretorius, I.S.; Thevelein, J.M.; Delvaux, F.R.

Glucose and sucrose: hazardous fast-food for industrial yeast? **Trends in Biotechnology**, v. 22, n. 10, p. 531-537, 2004.

Waclawovsky, A.J.; Sato, P.M.; Lembke, C.G.,; Moore, P.H.; Souza, G.M. Sugarcane for bioenergy production: an assessment of yield and regulation of sucrose content. **Plant Biotechnology Journal**, v. 8, n. 3, p. 263-276, 2010.

Walker, G.M. Yeast metabolism. In: WALKER, G.M. (Ed.). **Yeast Physiology and Biotechnology**. West Sussex: John Wiley & Sons Ltd, 1998, chap. 5, p. 203-255.

Walker, G.M. Yeast Nutrition. In: Walker, G.M. (Ed.). **Yeast Physiology and Biotechnology**. John Wiley & Sons, 1998, chap. 3, p. 51101.

Walter, A.; Rosillo-Calle, F.; Dolzan, P.; Piacente, E.; Cunha, K.B. Perspectives on fuel ethanol consumption and trade. **Biomass and Bioenergy**, v. 32, n. 8, p. 730-748, 2008.

Wang, L.; Zhao, X.Q.; Xue, C.; Bai, F.W. Impact of osmotic stress and ethanol inhibition in yeast cells on process oscillation associated with continuous very-high-gravity ethanol fermentation. **Biotechnology for Biofuels**, v. 6, n. 133, p. 1-10, 2013.

Wheals, A.E.; Basso, L.C.; Alves, D.M.G.; Amorim, H.V. Fuel ethanol after 25 years. **Trends in Biotechnology**, v. 17, n. 12, p. 482-487, 1999.

Wiame, J.M.; Grenson, M.; Arst, H.N. Nitrogen catabolite repression in yeasts and filamentous fungi. **Advances in Microbial Physiology**, v. 26, p. 1-88, 1985.

Winderickx, J.; De-Winde, J.H.; Crauwels, M.; Hino, A.; Hohmann, S.; Van-Dijck, P.; Thevelein, J.M. Regulation of genes encoding subunits of the trehalose synthase complex in *Saccharomyces cerevisiae*: novel variations of STRE-mediated transcription control? **Molecular Genetics and Genomics**, v. 252, n. 4, p. 470-482, 1996.

Wuebbles, D.J.; Jain, A.K. Concerns about climate change and the role of fossil fuel use. **Fuel Processing Technology**, v. 71, n. 1-3, p. 99119, 2001.

Yazawa, H.; Iwahashi, H.; Uemura, H. Disruption of URA7 and GAL6 improves the ethanol tolerance and fermentation capacity of *Saccharomyces cerevisiae*. **Yeast**, v. 24, p. 551-560, 2007.

Yoshikawa, K.; Tanaka, T.; Furusawa, C.; Nagahisa, K.; Hirasawa, T.; Shimizu, H. Comprehensive phenotypic analysis for identification of genes affecting growth under ethanol stress in Saccharomyces cerevisiae. **FEMS Yeast Research**, v. 9, n. 1, p. 32-44, 2009.

Zago, E.A.; Silva, F.L.F.; Bernardino, C.D.; Amorim, H.V. **Métodos analiticos para o controle da produçâo de etanol**. Piracicaba: FERMENTEC/FEALQ/ESALQ-USP, 1996. 194p.

Zakrajsek, T.; Raspor, P.; Jamnik, P. *Saccharomyces cerevisiae* in the stationary phase as a model organism - characterization at cellular and proteome level. **Journal of Proteomics**, v. 74, n. 12, p. 2837-2845, 2011.

Zaman, S.; Lippman, S.I.; Zhao, X.; Broach, J.R. How *Saccharomyces* responds to nutrients. **Annual Review of Genetics**, v. 42, p. 27-81, 2008.

Zhao, X.Q.; Bai, F.W. Mechanisms of yeast stress tolerance and its manipulation for efficient fuel ethanol production. **Journal of Biotechnology**, v. 144, n. 1, p. 23-30, 2009.

3 Potential of *Saccharomyces cerevisiae* over-expressing the *MSN2* or *TRP1* gene for high-alcohol fermentations simulating the industrial conditions of Brazilian distilleries

Summary

During the industrial production of ethanol, one of the limiting factors in the process is the different types of stress imposed on the yeast at the same time. Mainly the stress caused by the high concentrations of sugars present in the must and, after the fermentation process, by the high levels of ethanol produced by the yeast. Modification of the promoter region of the *TRP1* gene, involved in tryptophan synthesis, and of the *MSN2* gene, which encodes a transcription factor (Msn2) that regulates the general response to stress in *S. cerevisiae*, have been pointed out as attractive alternatives for increasing the tolerance of strains to the high alcohol content and other stresses faced in the Brazilian process. In this context, the aim of this study was to evaluate the potential of *S. cerevisiae strains* isogenic to the industrial strain CAT-1 with over-expression of the *MSN2* or *TRP1* genes for fermentations with high alcohol content and cell recycling, simulating the industrial conditions of Brazilian distilleries. Micro-scale growth tests were conducted by varying the concentrations of sugar and ethanol supplied. Fermentation experiments with cell recycling were conducted by varying the sugar concentrations and fermentation temperature. Over-expression of the *TRP1* gene for tryptophan synthesis favored the NAB-1 strain for greater growth in YEPD medium with 8% ethanol (v/v), compared to all the other strains tested. However, NAB-1 showed lower viability in growth tests in YEPD medium with 16 and 18% ethanol (v/v), and in molasses musts with 27 and 33% ART. NAB-1 also showed lower accumulation of reserve carbohydrates (trehalose and glycogen) in the fermentations with recycle. The results obtained revealed that the level of over-expression of the *MSN2* gene, or its truncated form, contributes to different physiological behavior between the ANT-5 and ATT-6 strains, since ANT-5 proved to be more sensitive, i.e. less viable than the CAT-1 and ATT-6 strains in treatments with 16 and 18% ethanol (v/v). ANT-5 behaved similarly to the parental strain in most of the parameters analyzed in fermentations with cell recycles. The ATT-6 strain, with the *MSN2* gene overexpressed in the truncated version, showed greater cell viability than the parental CAT-1 and the other strains in the microscale growth tests in musts containing 27 and 33% ART. The ATT-6 strain also showed higher ethanol production, fermentation speed and sugar consumption in fermentations with cell recycling where drastic increases in the must's sugar concentration were used. Among the strains evaluated, ATT-6 showed biotechnological potential to withstand multiple stresses encountered in fermentations with high alcohol content, under conditions typical of Brazilian distilleries.

Keywords: Alcoholic fermentation; Alcoholic stress; Osmotic stress; *Saccharomyces cerevisiae; TRP1 / MSN2* gene overexpression; General stress response

3.1 Introduction

In recent years, the importance attached to biofuels has grown worldwide, mainly due to global warming and the future depletion of oil (Demirbas, 2017). In addition, the prospecting and use of oil derivatives, such as diesel and gasoline, involves high costs and climate concerns (Marchal et al., 2006). Current market conditions make ethanol production a promising activity. The availability of fossil fuels will not be able to meet the growing demand for energy on our planet. Therefore, ethanol production will gradually expand (Debnath et al., 2017).

Ethanol is produced from different raw materials, classified into three categories: simple sugars, starch and lignocellulose (Balat & Balat, 2009). Sugarcane is a crop that can be used in tropical countries to produce ethanol, both in the form of juice and molasses for the preparation of musts. This raw material is mainly composed of sucrose, fiber and water, usually in proportions of approximately 12, 15 and 70%, respectively (Chen & Chou, 1993). The conversion of molasses or cane juice into ethanol is easier compared to starch compounds and lignocellulosic biomass, since prior hydrolysis of the raw material before fermentation is unnecessary because sucrose can be metabolized directly by yeasts (Waclawovsky et al., 2010).

In Brazil, the oldest and most widely used process in distilleries is Batch

Fermentation, where basically a yeast suspension is prepared at the bottom of the vat, and then the vats receive sugar cane molasses (must) on top. At the end of the must supply, the yeasts make up around 10% of the total content of the vat, carrying out the fermentation with a high cell density (Godoy et al., 2008). In the 1930s, the Melle-Boinot procedure was adopted, adding to the process: treatment of the yeast with water and acid, centrifugation and reinsertion of the yeast for sequential fermentation (> 90% of the cells are reused in the next fermentation cycle) (Boinot, 1939; Wheals et al., 1999; Basso et al., 2008).

Although Batch Fermentation has contributed to higher yields, due to the reuse of cells, in this procedure the yeasts face a series of stress factors. These factors impose restrictions on the cells, affecting their growth and metabolism (Basso et al., 2008). Among the stress factors, we can list: high sugar concentration, presence of salts in the must, high temperatures, high ethanol concentration, low extracellular pH, bacterial contamination, nutrient deficiency and the presence of inhibitory agents (Basso et al., 2011). In addition, cell recycling can cause a decrease in cell viability over several consecutive ci clos. On the other hand, this process has a positive aspect: during cell recycling, the selective pressures imposed by the stress factors select multi-tolerant strains capable of resisting this process in industrial fermentation (Basso et al., 2008).

The Brazilian process, together with *Very High Gravity Fermentation* (*VHG Fermentation*), can contribute to even higher ethanol yields. VHG Fermentation is characterized by obtaining high levels of alcohol using musts with more than 25% (w/v) ART. *VHG fermentation* can reduce the use of water in the process and consequently reduce the production of vinasse (Thomas et al., 1996). The use of this technology, in addition to saving water by approximately 40%, also reduces energy use because less fluid is sent to the heating, cooling and distillation stages (Puligundla et al., 2011). However, the use of this technology combined with the Brazilian process can cause additional deleterious effects on yeasts. Therefore, it is of fundamental importance to obtain robust and multi-tolerant strains, so that they can survive ethanolic stress, at the cost of a previous osmotic stress, withstanding all the stresses listed above for several consecutive cycles.

A promising way of obtaining multi-tolerant yeasts would be to select industrial strains already adapted to the Brazilian process and make genetic modifications to increase their tolerance even further. In *S. cerevisiae*, the transcription factors Hsf1 and Msn2/Msn4 activate genes containing stress responsive elements (STRE) (Smith et al., 1998). The functional STRE sequence contains 5 bp and is capable of activating the expression of genes in response to various stimuli harmful to cells (Nicholls et al., 2004), including oxidative stress, low nutrient availability, low pH, exposure to ethanol and osmotic stress (Kobayashi & Mcentee, 1991, 1993; Marchler et al., 1993; Schüler et al., 1994). Other studies have shown that there are correlations between amino acid biosynthesis and ethanol stress in yeast (Takagi et al., 2005; Hirasawa et al., 2007; Li et al., 2010). Over-expression of the *TRP1* and *TRP5* genes, which are involved in tryptophan biosynthesis, can result in greater tolerance to ethanol stress (Hirasawa et al., 2007). Therefore, genetic modification of the promoter region of tryptophan biosynthesis genes, or of the *MSN2* gene, which encodes a transcription factor (Msn2) that regulates the general response to stress in *S. cerevisiae,* could contribute to increasing the tolerance of strains to the different types of stress faced in industry and also in *VHG fermentation.*

In this context, the aim of this work was to evaluate the potential of genetically modified *S. cerevisiae* strains for fermentations with high alcohol content and cell recycling. To this end, strains with over-expression of the *MSN2* and *TRP1* genes were compared physiologically with the CAT-1 parental strain under conditions that simulate Brazilian distillery conditions.

3.2 Material and Methods

3.2.1 Biological material

The strains used in this study were provided by Prof. Dr. Boris U. Stambuk of the Federal University of Santa Catarina. These strains were modified using genomic engineering techniques to insert the strong, constitutive PADH1 promoter into the promoter region of the target gene (Table 1) (Bücker, 2014). The procedure was carried out using PCR techniques and integration of the overexpression module into the genome by homologous recombination. These techniques made it possible to modify the expression of the gene by inserting a nucleotide sequence that codes for a constitutive promoter region, which allows the over-expression of the gene of interest. The following were carried out: modification of the promoter region of the *TRP1* gene involved in tryptophan synthesis (NAB-1 strain), modification of the *MSN2 gene*, which encodes a transcription factor (Msn2) that regulates the general stress response in *S. cerevisiae (*ANT-5) and a truncated version of the *MSN2-T* gene, without the first 48 amino acids (ATT-6).

Gene expression analyses were carried out by qRT-PCR. RNA from the recombinant and parental strains was extracted and subjected to reverse transcription (Bücker, 2014). Quantifications revealed that the *TRP1* gene, over-expressed in the NAB-1 strain, is 20 times more expressed than in the parental CAT-1 strain. The over-expressed *MSN2* gene (ANT-5) and its truncated version (ATT-6) are 7 and 8 times more expressed than in the parental strain, respectively.

Table 1 - Genotypes of the modified strains and the parental strain CAT-1

Lineage	Genotype	Reference
CAT-1	Industrial diploid	Basso et al. (2008)
NAB-1	Isogenic to CAT-1 but *loxP-KanMX-loxP-PADH1::TRP1*	Bücker (2014)
ANT-5	Isogenic to CAT-1 but *loxP-KanMX-loxP-PADH1::MSN2*	Bücker (2014)
ATT-6	Isogenic to CAT-1 but *loxP-KanMX-loxP-PADH1::MSN2-T*	Bücker (2014)

3.2.2 Micro-scale growth tests in different concentrations of ART or ethanol

Growth tests were carried out in triplicate on a TECAN spectrophotometer (Optical Density/D.O. 600nm), by incubating 96-well microplates in microaerobic conditions with readings at 2-hour intervals at 30 °C for a period of 72 hours. Shaking was carried out for 5 minutes before each reading. To assemble the microplate, 10 µL of inoculum pre-grown for 48 hours in YEPD medium (2% D-glucose, 1% yeast extract and 1% bacteriological peptone) and 90 µL of cane molasses mash with different concentrations of ART (33, 30, 27, 25, 23 and 19%, w/v) or YEPD medium with different levels of ethanol (18, 16, 14, 12 and 8%, v/v) were added to each well. After 72 hours of growth, an aliquot of 50 µL was taken from each well for viability analysis using an optical microscope.

3.2.3 Preparation of substrate from sugar cane molasses

The substrates/musts from sugar cane molasses were prepared by diluting the molasses with water to obtain the desired percentage (%, w/v) of ART. The diluted substrate was centrifuged at 10,000 rpm under 20°C for 20 minutes and placed in erlenmeyer flasks, which were sterilized at 121°C for 25 minutes.

3.2.4 Propagation of yeast biomass for fermentation trials

For propagation, 200 µL of each culture previously stored in an ultra-freezer (in a 20% glycerol solution) were transferred to tubes containing 5 mL of YEPD medium and incubated at 28°C for 48 hours. After growth, the entire contents of each tube were transferred to erlenmeyer flasks containing 100 mL of molasses mash with 10% ART and pH 4.5, incubated at 30°C, 150 rpm, for 24 hours.

Subsequently, the total contents of each flask were transferred to flasks containing 1 L of the same must mentioned above. The flasks were incubated at 30°C, 100 rpm, for 48 hours.

In the first fermentation cycle, the wet biomass of each propagated strain was collected in conical bottom tubes (*Falcon* tubes with a capacity of 50 mL) by centrifugation at 4000 rpm. The biomasses were adjusted to start the first fermentation with the same yeast content.

3.2.5 Fermentation tests

The fermentation trials were carried out in triplicate, simulating the industrial conditions of Brazilian distilleries with cell recycling. For the inoculum of the first fermentation cycle, 3.0-3.5 g of biomass of each strain was collected by centrifugation. Next, the "vat foot" was prepared, in which the biomass of each strain was resuspended in 2 mL of the fermented contents of the propagation stage (de-oiled wine) and diluted in 6 mL of distilled water. Subsequently, 28 mL of molasses substrate (must) was added, divided into 3 equal portions, spaced 2 hours apart. The fermentations were conducted at 30 or 32°C. This procedure sought to simulate what happens in the industrial process, i.e. a vat stand with around 33% yeast suspended in a mixture of water and wine, with the vat stand representing around 25-30% of the final fermentation volume. The must was divided into 3 portions to avoid pure batch feeding, i.e. in a single addition.

At the end of the first fermentation cycle, the yeast biomass was separated from the fermented substrate by centrifugation (4000 rpm) and weighed. Next, a new "vat stand" was prepared, in which 2 mL of the wine removed from the previous fermentation was added (simulating a "yeast cream" with a concentration of ~ 70% from an industrial centrifuge). The "yeast cream" was diluted with 6 mL of distilled water, acidified with sulfuric acid to pH = 2.5-2.6, and kept for 1 hour in this condition to simulate the acid treatment of the yeast carried out in the industrial plant. From then on, the "feed" with 28 mL of must was divided into three equal portions, according to the conditions described above. The whole procedure was repeated successively, i.e. in 15 cycles in the first fermentation experiment and 7 cycles in the second. The ART concentration (%, w/v) of the must was gradually increased throughout the cycles in both fermentation experiments.

The ethanol content of the settled wine was measured in all cycles. The concentration of glycerol and residual sugars in the first and last cycles was also analyzed. In addition, the wet biomass content, cell viability and bacterial contamination were estimated in all cycles.

3.2.6 Quantification of residual sugars, glycerol and ethanol

The concentration of residual sugars (sucrose, glucose and fructose) and glycerol in the wines were determined by High Performance Liquid Chromatography (HPLC), using a Dionex DX-300 ion exchange chromatograph (Sunnyvale, CA, USA) equipped with a CarboPac PA-1 column (4 x 250 mm) and pulse amperometry detector. The mobile phase used was NaOH (100 mM) at a flow rate of 0.9 mL.min^{-1} (Basso et al., 2008).

To determine the alcohol content of the de-oaked wines, 10 or 25 mL of the wine was transferred into a Kjeldahl micro-distiller (Piracicaba, Brazil). After distillation by vapor drag, the condensed material was collected in a 50 mL volumetric flask until the volume was complete. The distilled samples were transferred to an Anton-Paar DMA-48 digital densimeter (Graz, Austria) to estimate the ethanol content (%, v/v). The value generated by the densimeter was multiplied by 5 or 2, when 10 or 25 mL of wine was distilled, respectively, in order to compensate for the dilution that occurred during the distillation stage.

3.2.7 Estimation of fermentation speed and determination of biomass

The tubes were weighed at two-hour intervals during each cycle. The fermentation rate refers to the release of CO_2 (g) over the hours.

Before starting the fermentations, the empty conical-bottomed tubes were weighed. At the end of each cycle, after separating the wines from their respective yeasts, the precipitated wet biomass was weighed (g). The wet biomass content at the end of each fermentation cycle was determined by weighing the difference in weight between the tubes with and without yeast.

3.2.8 Yeast cell viability and control of bacterial contamination

Cell viability was estimated by optical microscopy, under a 40x objective, by differential cell staining using a solution of erythrosine in a phosphate buffer. Viable unstained cells and non-viable pink-stained cells were counted in a *Neubauer* chamber. Viability was expressed as a percentage, based on the ratio of viable cells to the total number of cells counted (viable and non-viable). The number of viable bacteria (rod-shaped) was estimated by light microscopy using a 100x objective and a glass slide. An aliquot of 100 μL of the sample was previously placed in contact with 100 μL of a solution of methylene blue (0.2%) and nile sulphate (2%). In this method, non-viable rods are stained blue, while live rods remain colorless. Only live bacteria are counted and bacterial contamination is expressed in cells.mL^{-1} (Oliveira et al., 1996).

3.2.9 Extraction and quantification of reserve carbohydrates

The trehalose content in the yeast was estimated by differential extraction with trichloroacetic acid followed by chromatographic assay as described above (item 3.2.6) (Christofoleti-Furlan, 2012).

For the extraction of trehalose, 200 mg of wet biomass was washed with ice-cold distilled water and precipitated by centrifugation (3500 rpm for 6 minutes) in a conical-bottomed tube. Extraction was carried out by adding 2 mL of 0.5 M trichloroacetic acid in an ice bath for 20 minutes and stirring occasionally. It was then centrifuged again (3500 rpm for 6 minutes) and the supernatant extract was collected and stored at -20°C until quantification analysis by HPLC (Trevelyan & Harrison, 1956).

To extract the glycogen, a suspension of 200 mg of wet biomass was placed in test tubes with screw caps. The contents of the tubes were resuspended in 5 mL of ice-cold distilled water and the tubes were centrifuged at 3500 rpm for 6 minutes. The supernatant was discarded and 2 mL of 0.25 M sodium carbonate was added. The tubes were stored at -20°C until quantification analysis. Glycogen quantification was carried out by hydrolysis

enzymatic with amyloglucosidase and colorimetric measurement of glucose released after reaction with glucose oxidase and peroxidase. Quantification was determined using a spectrophotometer (500 nm) (Becker, 1978).

3.2.10   Statistical analysis

Statistical analyses were carried out using analysis of variance (ANOVA) and Tukey's mean comparison tests in the R software. The means for the triplicates of each strain were considered different at a 5% significance level ($p<0.05$).

3.3 Results and Discussion

This work was divided into four stages. In the first and second stages, microscale tests were carried out under microaerobic conditions to assess the growth and viability of the strains in musts/media containing different concentrations of sugar (item 3.3.1) and ethanol (item 3.3.2), respectively. In the third stage, a fermentation experiment was carried out, in which the biomass content of each strain was subjected to 15 consecutive fermentations, in which the sugar concentration of the sugar cane molasses must was gradually increased with each cell recycle (item 3.3.3). In the fourth and final stage, seven consecutive fermentations were carried out, with more drastic increases in the sugar concentrations in the cell cycles (item 3.3.4).

3.3.1 Growth and viability at different ART concentrations in sugar cane molasses must

In the first stage of this work, growth tests under microaerobic conditions were carried out on musts containing 33, 30, 27, 25, 23 and 19% ART. No significant differences were observed in the growth curves obtained by the four strains at the different concentrations. All the strains behaved equally as the sugar concentration in the must increased (Figure 1). At concentrations of 33, 30, 27 and 25% ART, the stationary phase was reached after 54, 48, 42 and 36 hours, respectively (Figure 1A, B, C and D). As expected, and also reported in the study by Munna et al. (2015), the higher the concentration of sugar supplied, the longer the strains took to reach stationary phase. This was possibly due to the greater amount of sugar available in the must and also due to the greater exposure of the cells to osmotic stress, allowing them to grow less and more slowly. Above 27% ART, it was clearly observed that the growth behavior changed, increasing the period of the logarithmic phase and considerably decreasing the final growth (D.O.) (Figure 1A, B, C). At concentrations of 23 and 19% ART, the strains reached the stationary phase after 36 and 30 hours, respectively (Figure 1E, F).

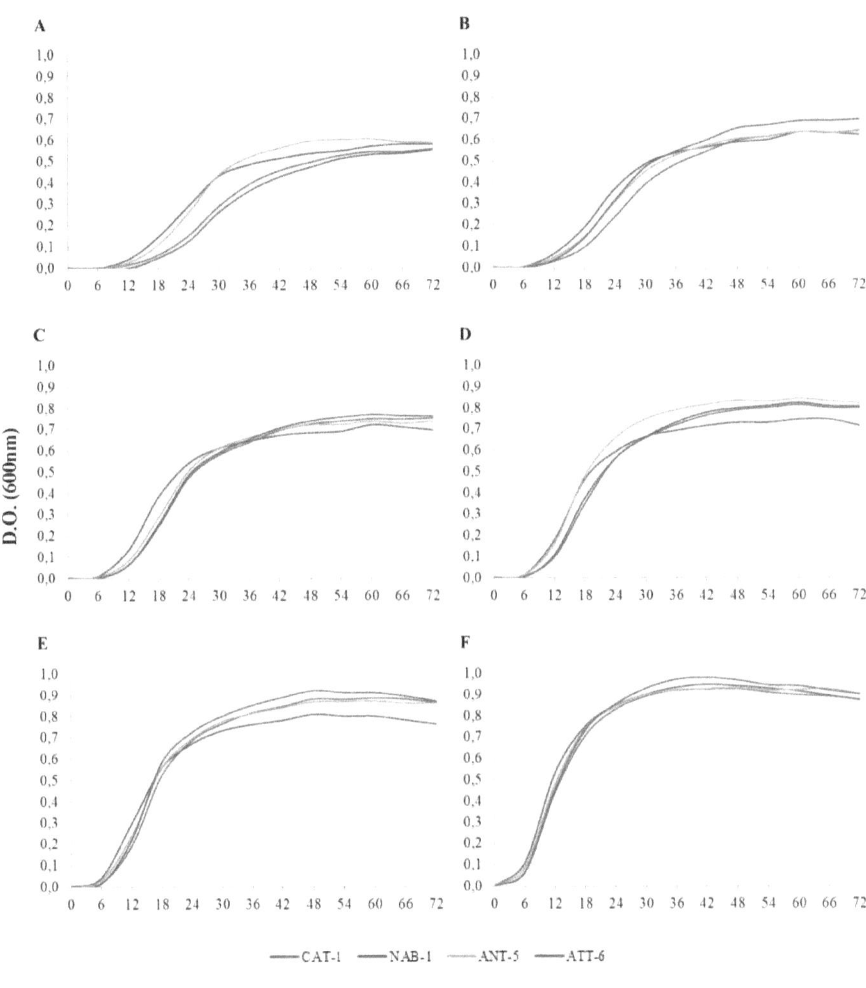

Tempo (horas)

Figure 1. Growth curves measured for strains CAT-1, NAB-1, ANT-5 and ATT-6 at different ART concentrations. A) 33%; B) 30%; C) 27%; D) 25%; E) 23%; F) 19% ART in sugar cane molasses must

The averages for the final O.D. (72 hours) of the four strains were: 0.887; 0.838; 0.784; 0.736; 0.650; and 0.570 for the concentrations of 19, 23, 25, 27, 30 and 33% ART, respectively (Figure 2A).

The final viability (72 hours) was analyzed by taking an aliquot from the microplate wells of the treatments with 23, 27 and 33% ART (Figure 2B). The strains showed no difference in viability at 23% ART. However, at concentrations of 27 and 33% ART, the ATT-6 strain differed significantly from the others, showing the highest viability values: 86.13 and 77.75%, respectively. The parental strain CAT-1 showed viability values of 77.24 and 69.78%, and did not differ from ANT-5, which showed 79.21 and 68.49% at these same ART concentrations (27 and 33%). The NAB-1 strain showed the lowest viability: 27.06 and 22.26%, at concentrations of 27 and 33% ART, respectively, differing from the other strains tested (Figure 2B).

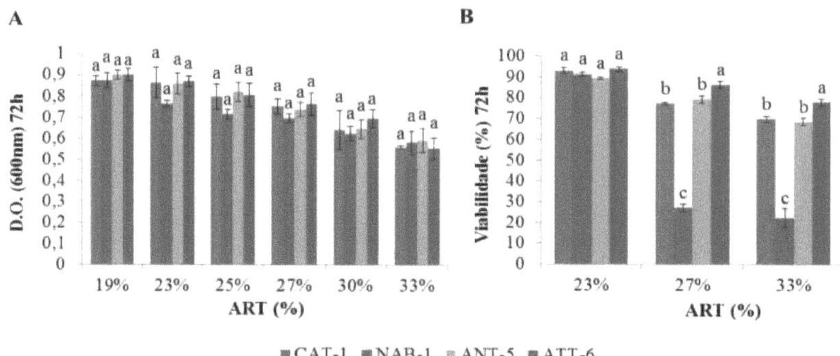

Figure 2. Final D.O. data (A), viability (B) and statistical analyses of the CAT-1, NAB-1, ANT-5 and ATT-6 strains at 72 hours in sugarcane molasses mash with different ART concentrations (%). For each treatment, equal letters do not differ at 5% significance level (p<0.05), according to Tukey's test.

The genetic modifications did not affect the growth of the strains at the different sugar concentrations (% ART), as the growth curves and final O.D. values were similar between the strains (Figure 1 and 2A). The cell viability tests proved important to complement the results obtained from the O.D. values, where it was not possible to observe differences between the strains.

The cell viability results indicated that ATT-6 (with Msn2 transcription factor over-expressed in the truncated version) showed significantly higher values than the other strains, both in must with 27% ART (86.13% viability) and at 33% ART (77.75% viability) (Figure 2B).

Exposure of cells to media containing 20% or more ART triggers the stress response mechanism in *S. cerevisiae* (Gomar-Alba et al., 2012). In our experiments, over-expression of the truncated version of the *MSN2* gene increased cell tolerance to osmotic stress caused by the high sugar concentrations present in the must (Figure 2B). According to Gasch et al. (2000), the Msn2 and Msn4 transcription factors induce the transcription of genes whose promoters contain the stress response element (STRE) during a variety of stress conditions, including osmotic stress. The HOG1 pathway, which is activated in response to an increase in external osmolarity, and the cAMP-PKA pathway, involved in sensing the nutritional state of the cell, have been described as capable of playing regulatory roles in the control of *MSN2/4* (Marchler et al., 1993; Schüller et al., 1994). According to Alepuz et al. (2001), the Msn2 and/or Msn4 transcription factors recruit Hog1 to the *CTT1* gene promoter (cytosolic catalase synthesis), suggesting a functional interaction between the HOG1, Msn2 and Msn4 pathway. The opposite was observed for the NAB-1 strain, which behaved with lower viability in these same treatments (27 and 33% ART) compared to the parental strain, inferring that over-expression of the gene

TRP1, involved in tryptophan synthesis, did not confer tolerance to the cells when subjected to osmotic stress. The ANT-5 strain showed similar viability to the parental strain, demonstrating that possibly the level of over-expression of the *MSN2* gene was not enough to result in significant differences in the treatments with sugar cane molasses musts at concentrations of 27 and 33% ART.

3.3.2 Growth and viability in YEPD medium with different concentrations of ethanol

In the second stage, tests were carried out to assess the growth of the strains in YEPD medium containing 18, 16, 14, 12, 8 and 0% ethanol (v/v). At this stage, different growth behaviors were observed between the strains (Figure 3). At the highest concentration - 18% ethanol (v/v) - the parental strain CAT-1 achieved the greatest growth and reached stationary phase after 36 hours (Figure 2A). At 16% ethanol (v/v), the CAT-1 and ATT-6 strains reached stationary phase after 36 and 30 hours (Figure 3B), respectively. The same pattern was observed at 14% (v/v), where the CAT-1 and ATT-6 strains showed a similar growth profile and reached stationary phase after 18 hours, unlike NAB-1 and ANT-5, which reached stationary phase after 24 hours (Figure 3C). Under an ethanol concentration of 12% (v/v), the strains did not differ significantly and stationary phase occurred after 18 hours (Figure 3D). At a concentration of 8% (v/v) ethanol, the strains reached stationary phase after 12 hours, and NAB-1 grew significantly more than the others (Figure 3E). In YEPD medium without ethanol, used as a control, NAB-1 and ATT-5 showed a similar growth profile, with greater growth compared to the other strains (Figure 3F). On the other hand, CAT-1 and ATT-6 showed a similar growth profile, with less growth in the control medium (Figure 3F). Without the addition of ethanol to the medium, all the strains reached stationary phase quickly, after 6 hours of growth (Figure 3F).

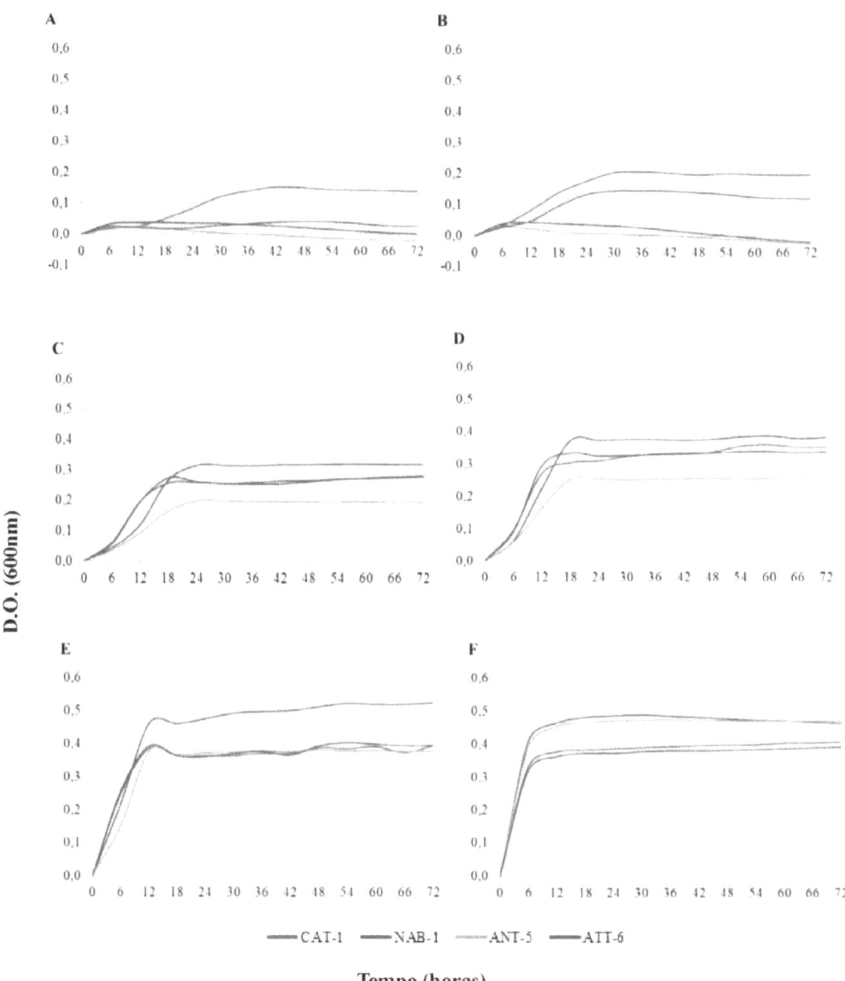

Figure 3: Growth curves measured for the CAT-1, NAB-1, ANT-5 and ATT-6 strains at different ethanol concentrations (%, v/v). A) 18%; B) 16%; C) 14%; D) 12%; E) 8%; F) 0% ethanol (v/v) in YEPD medium.

The final O.D. results, verified after 72 hours, revealed that the CAT-1 and ATT-6 strains showed similar growth in all ethanol treatments (Figure 4A). In the control medium, without the addition of ethanol, NAB-1 and ANT-5 (0.463 and 0.472) showed similar O.D. values, and higher than CAT-1 and ATT-6 (0.408 and 0.391). At 8% ethanol (v/v), the NAB-1 strain showed greater growth (0.524) than the other

53

strains tested over the 72-hour period. At concentrations of 12 and 14% ethanol (v/v), the CAT-1 (0.352 and 0.278), NAB-1 (0.382 and 0.318) and ATT-6 (0.337 and 0.282) strains were not significantly different from each other, and grew significantly more than ANT-5 (0.256 and 0.194). In the media with the highest levels, 16 and 18% ethanol (v/v), only the CAT-1 (0.197 and 0.137) and ATT-6 (0.119 and 0.040) strains showed growth (Figure 4A).

In terms of final viability, the strains did not differ in YEPD medium without the addition of ethanol, and showed approximately 95% viability (Figure 4B). The NAB-1 strain, despite achieving greater growth in the treatment with 8% ethanol (v/v), showed lower viability (90.85%). In the treatments with 8 and 12% ethanol (v/v), CAT-1, ANT-5 and ATT-6 were no different, with approximately 95% viability. At 16% ethanol (v/v), the CAT-1 strain showed 92.39% viability and was not significantly different from ATT-6 (88.62%), both of which showed significantly higher viability values than the others. At the same concentration of ethanol (16%), NAB-1 showed significantly lower viability than strain ANT-5 (22.21 and 49.40%, respectively).

In the treatment with the highest concentration of ethanol (18% v/v), CAT-1 (84.62% viability) and ATT-6 (67.32% viability) had the highest viability values, with CAT-1 having the highest value. In this more concentrated treatment (18% v/v), NAB-1 (20.47% viability) and ANT-5 (30.69% viability) did not differ from each other (Figure 4B).

Our data corroborate the results shown by Sasano et al. (2012), which indicate that overexpression of the *MSN2* gene could increase the sensitivity of cells to high concentrations of ethanol, since in this study the ANT-5 and ATT-6 strains showed lower viability than the parental strain (CAT-1) in the treatment with the highest concentration of ethanol, 18% (v/v). On the other hand, with 16% ethanol (v/v), of the two strains with the *MSN2* gene overexpressed, only ANT-5 (with a lower level of overexpression than ATT-6) showed lower viability compared to the parental strain. This may indicate that the level of overexpression could be related to the greater tolerance of the cells, since the ATT-6 strain, with the Msn2 transcription factor overexpressed in truncated form, showed viability equal to the parental strain at this ethanol content. For a more accurate answer, a study on the level of over-expression of the *MSN2* gene is necessary, testing different gene expression promoters. As well as testing the expression of the *MSN2* gene in its truncated form with its natural promoter. Such a study could answer the question of whether increasing the level of over-expression could lead to a change in cell tolerance to high levels of ethanol.

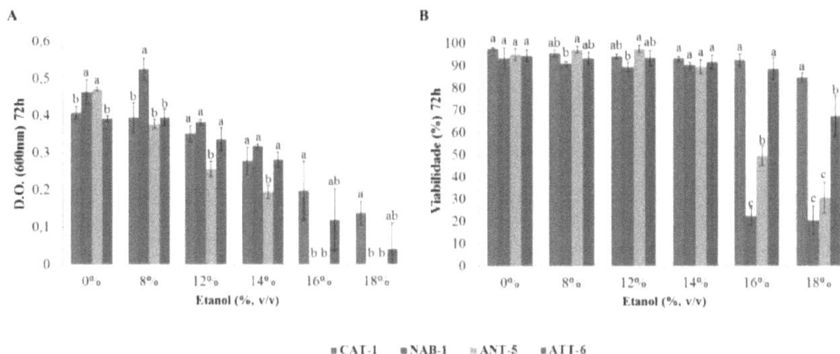

Figure 4 - Final OD data (A), viability (B) and statistical analysis of CAT-1, NAB-1, ANT-5 and ATT-6 strains at 72 hours in YEPD medium with different concentrations of ethanol (%, v/v). For each treatment, equal letters do not differ at 5% significance level (p<0.05), according to Tukey's test.

In summary, the treatments with different concentrations of ethanol affected the growth of the strains quite differently (Figure 3 and 4A). The NAB-1 strain showed significantly more growth (at 72 hours) than the other three strains in must containing 8% ethanol (v/v) (Figure 4A). However, NAB-1 did not grow when the ethanol concentrations were higher, 16 and 18% (v/v). The ATT-6 strain did not differ from the CAT-1 parental strain in terms of growth in the different ethanol treatments. Viability analyses revealed that the ATT-6 and CAT-1 strains had the highest values (Figure 4B). The ATT-6 strain only differed from the parental strain in the 18% (v/v) ethanol treatment, as it showed significantly lower viability. Therefore, in terms of growth and viability in the ethanol treatments, the ATT-6 strain was similar to the parental CAT-1, where the only difference was observed in viability with 18% (v/v) ethanol (84.63 and 67.33% viability for CAT-1 and ATT-6, respectively) (Figure 4A, B). With regard to the four strains tested, the lowest viability values were observed in the treatments with 16 and 18% (v/v), where NAB-1 showed the lowest values, followed by ANT-5. Bearing in mind that the NAB-1 and ANT-5 strains did not grow in the treatments with 16 and 18% ethanol (v/v), the lower viability obtained by them is justified, since the same inoculated cell content withstood the high concentrations of ethanol during the 72-hour period, unlike CAT-1 and ATT-6, which resisted these treatments and showed growth. The data indicate that overexpression of the truncated form of the *MSN2* gene (ATT-6) did not show greater tolerance to ethanol in microscale tests. However, it also did not prove to affect cell growth and viability, as was observed in the ANT-5 strain (over-expressing the normal form of the *MSN2* gene), which had impaired growth and viability compared to the parental strain (CAT-1). This suggests an important role not only for the ability to over-express the *MSN2* gene, but also for the expressed form of the Msn2p transcription factor.

3.3.3 First fermentation experiment with 15 cell cycles

In the third stage of the work, 15 fermentation cycles were carried out. Here, the ART concentrations (%, w/v) of the musts were gradually increased. The ART concentration of the must used in the first cycle was considerably low (14.60%), in order to get the strains acclimatized and start the fermentation with a good performance, especially with high viability (Table 2). The must's ART concentration was gradually increased to allow the strains to adapt better to each recycle, avoiding osmotic shock and favoring them when they were subjected to higher and more stressful sugar concentrations. Furthermore, in order to differentiate fermentation performance between the strains, it was necessary to carry out a slow and gradual increase in sugar concentrations. Fermentation time and speed data are shown in Supplementary Figure 1.

3.3.3.1   Ethanol production, biomass accumulation and cell viability

The first 10 cycles were conducted using molasses from Usina Sao Manoel

(Açucareira Sao Manoel S/A, Sao Manoel - SP) to prepare the musts. In the last 5 cycles, molasses from the Furlan Mill (Usina Açucareira Furlan S/A, Santa Bàrbara D'Oeste - SP) was used. According to the results shown in Table 2, ethanol was produced as expected throughout the cycles as the ART content was increased. After changing the mash, in the eleventh cycle, there was a decrease in ethanol production, possibly related to the lower sugar concentration supplied in this cycle. The aim of changing the molasses was to induce conditions that would cause disparities between the strains, since in the previous cycles up to the eleventh, differences were only observed in cycles 5 and 10. In addition, the molasses from Usina Furlan is more impure than the molasses from Usina Sao Manoel.

In cycle 5, the concentration of ART supplied was 25%, and the NAB-1 strain was significantly different from the CAT-1 parental strain, with lower ethanol production (Table 2). In cycle 10 (31.57% ART), the NAB-1 strain differed from all the strains with lower ethanol production. In cycle 14, at a concentration of 30% ART, strain ANT-5 showed significantly higher ethanol production (13.33% v/v) than the parental CAT-1 (13.15%). On the other hand, in the last cycle, with a sugar concentration of 33.33%, the ethanol production of the parental strain CAT-1 was significantly higher (15.13% v/v) compared to ANT-5 (14.68% v/v) and ATT-6 (14.65% v/v), but did not differ from NAB-1 (14.80% v/v) (Table 2).

Table 2 - Concentration of ART supplied, ethanol produced and fermentation temperature in the 15 cycles

Cycle	ART (%) of must	Ethanol (%,v/v) final media[1]				Temperature
		CAT-1	NAB-1	ANT-5	ATT-6	
1	14	7,03[a]	7,00[a]	7,00[a]	7,00[a]	30°C
2	18	8,92[a]	8,82[a]	8,88[a]	8,87[a]	30°C
3	20	9,95[a]	9,92[a]	9,98[a]	10,05[a]	30°C
4	22	10,95[a]	10,93[a]	11,12[a]	10,98[a]	30°C
5	25	12,17[a]	12,02b	12.05 b[a]	12.03 b[a]	30°C
6	27	13,32[a]	13,32[a]	13,35[a]	13,23[a]	30°C
7	29	13,75[a]	13,67[a]	13,70[a]	13,68[a]	30°C
8	31	15,07[a]	14,97[a]	15,17[a]	15,13[a]	30°C

9	31		15,03[a]	14,87[a]	15,03[a]	15,05[a]	32°C
10	31		14,17[a]	12,57b	14,35[a]	14,90[a]	32°C
11	20		10,85[a]	10,57[a]	10,57[a]	10,77[a]	32°C
12	20		9,40[a]	9,47[a]	9,45[a]	9,47a	32°C
13	25		11,07[a]	11,04[a]	11,23[a]	11,23[a]	32°C
14	30		13,15b	13.33 b[a]	13,58[a]	13.50 b[a]	32°C
15	33		15,13[a]	14.80 b[a]	14,68b	14,65b	32°C

[1] *Equal letters within the same cycle do not differ at 5% significance level (p<0.05)*

The yeasts were adjusted to start the first fermentation with the same cell content (3.5 g). Until the fourth cycle, ATT-6 showed significantly lower biomass values (Figure 5). In the following cycles, ATT-6 was no different compared to the other strains. However, in cycle 14 it changed its behavior and differed from all the others, showing significantly higher biomass production (5.77 g) (Figure 5). In cycle 15, the ATT-6 strain did not differ from the CAT-1 parental strain, but both showed significantly higher biomass accumulations compared to the other two strains evaluated. The biomass content values, calculated by subtracting the total biomass value obtained in cycle 15 from the content value of the first cycle (3.5 g), were not similar between the strains: CAT-1 (1.44 g), NAB-1 (0.99 g), ANT-5 (1.02 g) and ATT-6 (2.12 g). In other words, at the end of all the fermentations, ATT-6 had the highest cell accumulation, followed by the parental CAT-1.

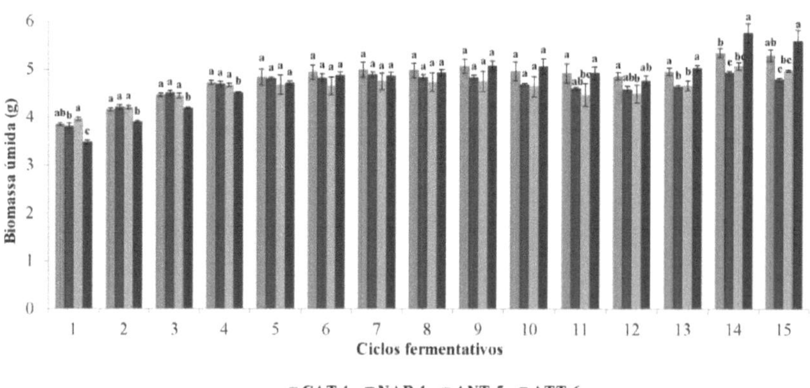

Figura 5. **Wet biomass production (g) and statistical analysis of CAT-1, NAB-1, ANT-5 and ATT-6 strains in 15 fermentation cycles. Equal letters within the same cycle do not differ at 5% significance level, according to Tukey's test.**

No bacterial contamination was detected in the fermentations, since the yeast was acid-treated from the second cycle onwards. During the cycles, the musts used had a small pH variation (5.3-5.5). For all the strains evaluated, cell viability remained stable at approximately 100% until the eighth cycle. Even after this cycle, it was observed that viability decreased minimally (Figure 6). A greater decrease in cell viability was expected, considering that the ART concentrations used in the recycles were high, especially in cycles 7, 8, 9, 10, 14 and 15 (Table 2). In addition, the stressful effect caused by the levels of ethanol produced by the strains, further intensified when the temperature was increased from 30 to 32°C (Table 2), could result in a greater reduction in cell viability. However, these conditions were not enough to cause a considerable decrease in viability, since even after the last cycle, the strains remained viable, with around 8090% of cells viable. One explanation for this behavior could be that the increase in sugar concentration (ART) from one cycle to the next was subtle, allowing the strains to adapt better, favoring them in the stressful situation encountered in each subsequent cycle. Some studies report that, in yeasts, the ability to cope with a more severe shock, from a different type of stress, can be significantly increased when the cells are exposed to a mild form of stress (Lewis et al., 1995; Park et al., 1997). In the last cycle analyzed, the NAB-1 strain, with an over-expressed *TRP1* tryptophan synthesis gene, and the CAT-1 parental strain, showed equal viability and were more viable than the other strains tested (Figure 6). However, these data did not allow us to say whether the strains tested here could be candidates for the process, since in most of the cycles analyzed, they showed a similar pattern of behavior. Therefore, at this stage of the work, we found no substantial differences between the strains in the main fermentation parameters analyzed: ethanol, biomass and viability. For this reason, we hypothesize that there were no differences between the modified strains and the parental one due to the way in which the experiment was conducted.

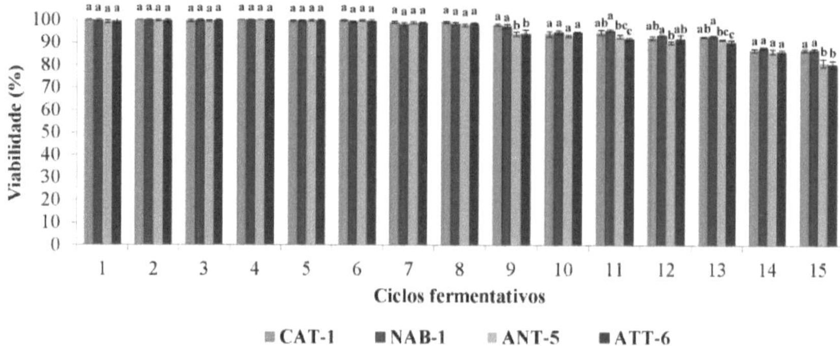

Figura 6. Cell viability (%) and statistical analysis of the CAT-1, NAB-1, ANT-5 and ATT-6 strains in the 15 fermentation cycles. Equal letters within the same cycle do not differ at 5% significance level, according to Tukey's test.

3.3.4 Second fermentation experiment with 7 cell cycles

In the last step of this work, seven fermentations were carried out using molasses from Usina Sao Manoel (Açucareira Sao Manoel S/A, Sao Manoel - SP). In this case, unlike the previous stage, the sugar concentration (ART) was drastically increased from one cycle to the next (Table 3). The aim was to impose more radical stress conditions, theoretically causing a greater deleterious effect on the strains and, consequently, increasing the possibility of detecting physiological differences in the fermentation parameters analyzed. No bacterial contamination was detected during the fermentations and the musts used had a pH range of 5.1-5.4 during the cycles.

Table 3 - Concentration of ART supplied and fermentation temperature in the 7 cycles

Cycle	ART (%) of must	Temperature
1	18	30°C
2	22	30°C
3	28	30°C
4	33	30°C
5	33	32°C
6	35	32°C
7	35	32°C

At this stage of the work, the strains were again adjusted to a similar biomass (3 g), with the aim of starting the first cycle under similar conditions. In this experiment, the CAT-1, NAB-1, ANT-5 and ATT-6 strains showed little growth, ending all cycles with 3.39; 3.05; 3.12; and 3.17 g of wet biomass, respectively. The lower cell growth obtained by the strains may be explained by the number of cycles, which was lower than the previous experiment (third stage).

3.3.4.1 Ethanol production, fermentation speed and cell viability

Regarding ethanol production (% v/v), differences were already observed from the first cycle, in which the biomass contained in the propagation medium with 10% ART was inoculated in the first fermentation cycle with fermentation must at a concentration of 18% ART. From this radical change in sugar concentration from the first fermentation, the ATT-6 strain was the highest producer of ethanol (% v/v) in cycles 1 (8.83), 2 (11.13), 3 (13.97), 4 (15.87) and 5 (15.08) ($p<0.05$) (Figure 7). In cycle 5, even after the temperature was changed to 32°C (Table 3) to cause an even more stressful condition, ATT-6 showed higher ethanol production. In cycles 6 and 7, ATT-6 produced 13.00 and 12.42% ethanol (v/v) and did not differ from strains CAT-1 (13.57 and 12.00%) and ANT-5 (13.07 and 11.97%). The ANT-5 strain showed similar behavior to the CAT-1 parental in terms of ethanol production in all 7 cycles. On the other hand, NAB-1 showed significantly lower ethanol production values compared to all the other strains, up until the seventh cycle (Figure 7).

It was possible to correlate the ethanol production data with the CO_2 release results, since from cycle 1 to 5 the ATT-6 strain showed the highest ethanol production (Figure 7) and also the highest CO_2 release (Figure 8). As a result, ATT-6 also showed a higher fermentation speed in cycles 1, 2, 4 and 5, i.e. in most of the cycles carried out (Figure 8).

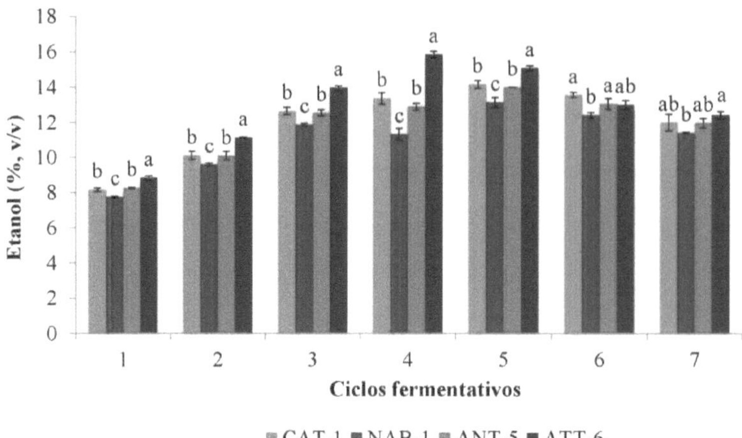

Figura 7.   Ethanol production (% v/v) and statistical analysis of CAT-1, NAB-1, ANT-5 and ATT-6 strains in the seven fermentation cycles. Equal letters within the same cycle do not differ at 5% significance level, according to Tukey's test.

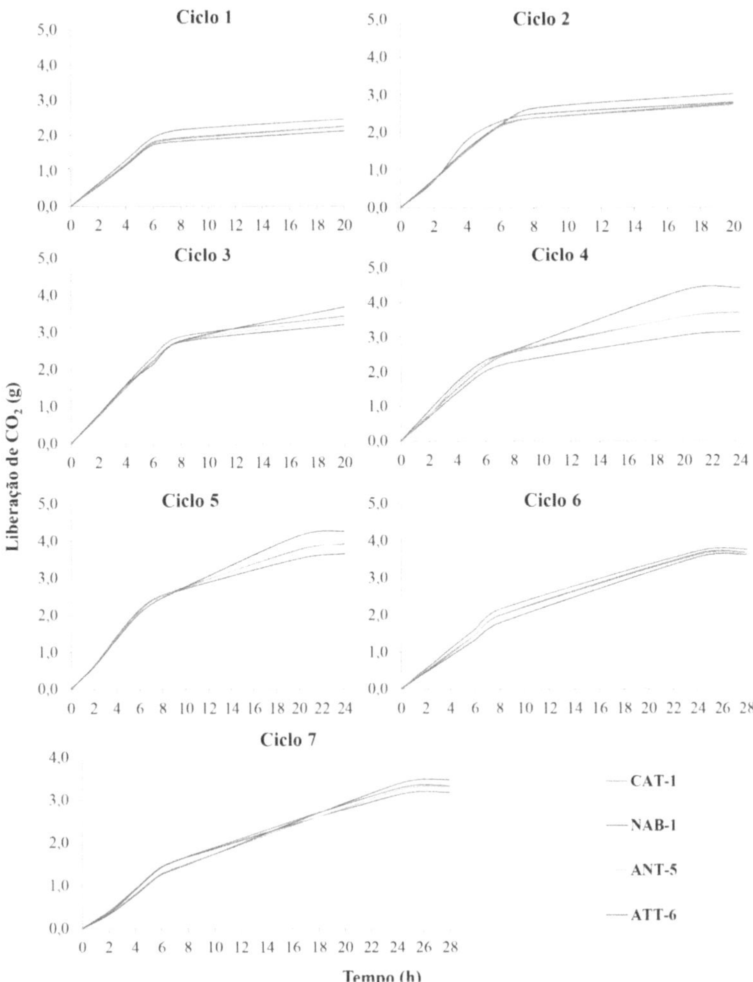

Figura 8. Fermentation speed of CAT-1, NAB-1, ANT-5 and ATT-6 strains, calculated by CO_2 released (g) over the hours, in the seven fermentation cycles

The behavioral change in CO_2 release observed in the ATT-6 strain from cycle 6 onwards can be explained by looking at the viability data (Figure 9). The ATT-6 strain showed a considerable decrease in viability at the end of cycle 5 (from 93.56% in the fourth cycle to 72.89% in the fifth), possibly due to the high concentrations of ethanol produced by this strain in previous cycles (Figure 7). In addition, the stress caused by ethanol may have been intensified by the increase in temperature in cycle

5, from 30°C to 32°C (Table 3). The higher the temperature, the higher the maximum intracellular alcohol concentration and, consequently, the more profound the inhibitory effect of ethanol (Navarro & Durand, 1978). Therefore, ATT-6 started fermentation in cycle 6 with lower viability, which may have contributed to the decrease in fermentation speed seen in this cycle (Figure 8).

Up to cycle 3, all the strains remained viable at around 98-99%. In cycles 4 and 5, for strain ATT-6, the viability data correlated with ethanol production (%, v/v), since ATT-6 (93.56% and 72.89%) showed significantly lower viability values compared to the parental strain CAT-1 (99.08% and 84.60%), and in the same cycles (4 and 5), ATT-6 showed the highest ethanol production values (Figure 7 and 9). It is worth noting that in the last two cycles, where the sugar concentration (% of ART) was increased to 35%, the ATT-6 strain maintained stable viability, unlike the CAT-1 parental strain, which showed a considerable drop in viability in cycle 7 (from 76.11% in cycle 6 to 65.33 in cycle 7) (Figure 9). The fact that the ATT-6 strain maintained stable viability at the highest concentration of ART provided in this experiment (35%) may reinforce the hypothesis that the *MSN2* gene in its truncated version may have favored the ATT-6 strain under osmotic stress.

There were no significant differences in viability between the strains in the seventh cycle. The CAT-1 parental strain stood out in terms of cell viability in cycle 6, where it obtained significantly higher viability (76.11%) compared to the other 3 strains analyzed. However, the viability results do not suggest that the CAT-1 strain is more tolerant to ethanol than ATT-6, because the parental strain produced lower levels of ethanol than ATT-6 in most of the cycles (cycles 1, 2, 3, 4, 5 and 7). In this way, the cellular content of CAT-1 may have been less affected than that of the ATT-6 strain. Therefore, from the results obtained in this experiment, we cannot conclude that the CAT-1 strain is more tolerant to ethanol than the ATT-6 strain, as the concentrations produced by the strains were not equivalent.

On the other hand, compared to the NAB-1 strain, the parental strain stood out with greater viability in this same cycle (cycle 6). This suggests that the NAB-1 strain, with the over-expressed tryptophan synthesis gene, is less tolerant to the effect of ethanol than the CAT-1 parental strain, since although it produced a lower ethanol content in cycle 6 (Figure 7), it showed lower viability (65.06%) (Figure 9). The same is true of the ANT-5 strain, which produced the same amount of ethanol as the parental strain in cycle 6, but also showed lower viability (63.53%) (Figures 7 and 9).

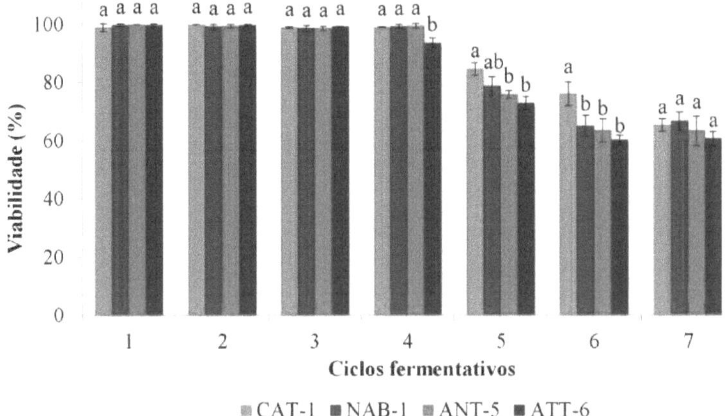

Figure 9. Cell viability (%) and statistical analysis of strains CAT-1, NAB-1, ANT-5 and ATT-6 in the 7 fermentation cycles. Equal letters within the same cycle do not differ at 5% significance level, according to Tukey's test.

3.3.4.2 Residual sugars and glycerol production

The residual sugar data quantified in ci clos 2, 3, 4 and 7 showed that ATT-6 was significantly different from the other strains evaluated because it had the lowest sugar values at the end of all the cycles analyzed (Figure 10). This result shows that ATT-6 consumed more sugar, which correlates with the ethanol production results (Figure 7).

In general, the fourth stage of this work revealed that ATT-6 showed the highest ethanol values in most of the recycles and also showed the highest fermentation speed. These data are correlated with higher sugar consumption, suggesting that this strain preferentially diverted the carbon source to obtain ethanol. In cycles 4 and 7, NAB-1 showed the highest residual sugar values and differed from the parental strain CAT-1 (Figure 10).

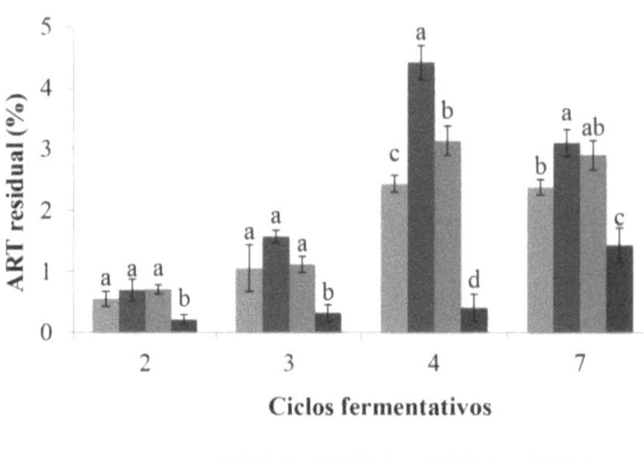

Figure 10. Total residual sugar (%) and statistical analysis of CAT-1, NAB-1, ANT-5 and ATT-6 strains at the end of cycles 2, 3, 4 and 7. Equal letters within the same cycle do not differ at 5% significance level, according to Tukey's test.

Glycerol was quantified in cycles 2, 3, 4 and 7 (Figure 11). In cycles 2 and 3, the glycerol production obtained by CAT-1 (0.50 and 0.68%) and ATT-6 (0.59 and 0.78%) was not different and both produced more glycerol than NAB-1 (0.42 and 0.63%) and ANT-5 (0.56 and 0.63%). In the fourth cycle, in which the must concentration was 33%, the ATT-6 strain showed higher glycerol production (0.78%) and, in this same cycle, this strain showed significantly higher ethanol production than the other strains tested (Figure 7). Glycerol and organic acids, together with biomass, are by-products related to the oxidation-reduction equilibrium under anaerobic conditions, since alcoholic fermentation is a non-oxidative process, and in order to maintain the oxidation-reduction equilibrium, all the NADH formed in oxidation reactions must be consumed in reduction reactions (coupled with the production of ethanol and glycerol) (Van-Dijken & Scheffers, 1986). Therefore, glycerol is the secondary product formed in the same pathway in which ethanol is synthesized. It is an osmoregulatory metabolite and its formation increases when there is a high osmotic pressure in the medium, protecting the cells in the presence of this stress (Guo et al., 2001, Cronwright et al., 2002). *S. cerevisiae* yeasts, when under osmotic stress, can trigger the HOG (*high osmolarity glycerol*) pathway. In this metabolic pathway, membrane receptors trigger a signaling cascade involving proteins that are transported to the nucleus and bind to promoters that can activate transcription factors involved in glycerol biosynthesis. By synthesizing glycerol, the HOG pathway is known to confer impermeability to cells (Hersen et al., 2008; Hohmann, 2002; Tatebayashi et al., 2006). When in conditions with high sugar concentrations, cells naturally lose water in order for osmotic regulation to occur. The loss of water causes an increase in the concentrations of ions and intracellular biomolecules, resulting in a decrease in cellular activity. The ability of yeast to withstand osmotic stress is directly related to the production and retention of glycerol (Saito & Posas, 2012).

According to Rep et al. (2000), the nuclear activity of the Msn2 and Msn4 transcription factors can be controlled by the Hog1 protein after a situation of osmotic shock. Global gene expression analysis showed an apparent correlation between Msn2/Msn4 and Hog1. Taking this information into account, it can be inferred that in cycle 4, the ATT-6 strain reacted better to the osmotic stress, because not only did it show a lower amount of residual sugar in this cycle (Figure 10), i.e. it had a higher sugar consumption, but it also showed a higher accumulation of ethanol (Figure 7) and glycerol (Figure 11). In cycle 7, CAT-1 produced 0.85% glycerol and was no different from NAB-1 (0.87%) and ATT-6 (0.77%), but was significantly different from ANT-5 (0.74%).

The data obtained here corroborate the information that glycerol can act as a regulator of osmotic stress in yeast, because as the sugar concentration (ART) was increased in cycles 2 (22%), 3 (28%), 4 (33%) and 7 (35%) (Table 3), there was also a corresponding increase in glycerol biosynthesis obtained by the four strains evaluated (Figure 11). In this sense, the over-expression of the truncated *MSN2* gene suggests that the HOG1 pathway is favored for glycerol formation, and consequently better osmotic control under the stress conditions imposed by the experiments with substantial increases in sugar concentration (ART).

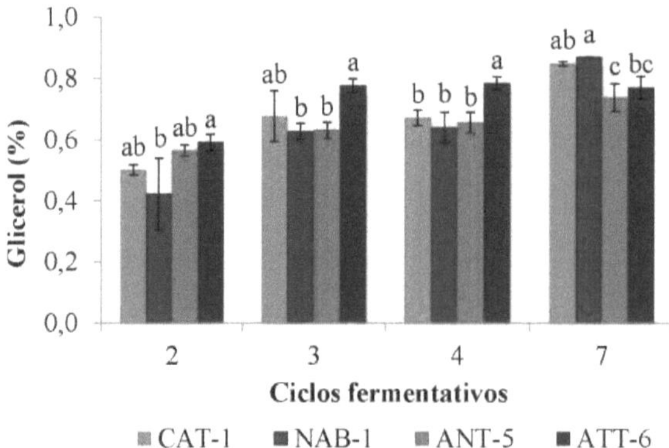

Figure 11. Glycerol biosynthesis (%) and statistical analysis of CAT-1, NAB-1, ANT-5 and ATT-6 strains at the end of cycles 2, 3, 4 and 7. Equal letters within the same cycle do not differ at 5% significance level, according to Tukey's test.

3.3.4.3 Reserve carbohydrates

With regard to reserve carbohydrates, analyzed at the end of the last cycle (%, mg/100mg of dry biomass), the trehalose values revealed that CAT-1 (4.82), ANT-5 (4.72) and ATT-6 (4.05) did not differ, but showed significantly higher trehalose accumulation values compared to the NAB-1 strain (2.91) (Figure 12).

Strain NAB-1 also showed lower glycogen accumulation and differed significantly from the other strains. The CAT-1, ANT-5 and ATT-6 strains did not differ in terms of glycogen production (Figure 13). In yeasts, osmotic stress and the consequent loss of water can cause different mechanisms related to the synthesis of glycerol and trehalose. In this way, cells can prevent dehydration and protect their structures (Mager & Siderius, 2002).

The reserve carbohydrates, trehalose and glycogen, can account for up to 30% of yeast dry matter. Trehalose, a disaccharide made up of two glucose molecules, is responsible for maintaining viable vegetative cells and spores. The protective function of trehalose is the result of the relationship between its location and cellular mobilization. It accumulates in the cytoplasm, influencing the water activity of the cytosol, contributing to cell latency and increasing resistance to water stress (Parrou et al., 1997; Singer & Lindquist, 1998; Zhao & Bai, 2009). In conditions where there is low water activity, trehalose replaces the water molecules in the membrane, on both sides of the phospholipid layer, maintaining its integrity. The levels of trehalose and glycogen fluctuate during fermentation and can end the process with different levels. These reserves can be metabolized by the cells at the end of fermentation, producing more ethanol and enriching the yeast with nitrogen (Amorim et al., 1996). The accumulation of trehalose is related to cell viability and tolerance to ethanol stress during fermentation (Mansure et al., 1997). According to Parrou et al. (1999), glycogen is the main reserve carbohydrate in yeasts. In conditions where nitrogen is limited, glycogen begins to accumulate in the cells immediately when the nitrogen source is exhausted. It is possible to hypothesize that the NAB-1 strain showed less glycogen accumulation because it over-expressed tryptophan constitutively. The over-expression of the *TRP1* gene in strain NAB-1 conferred a lower capacity to accumulate both reserve carbohydrates, compared to the other strains analyzed.

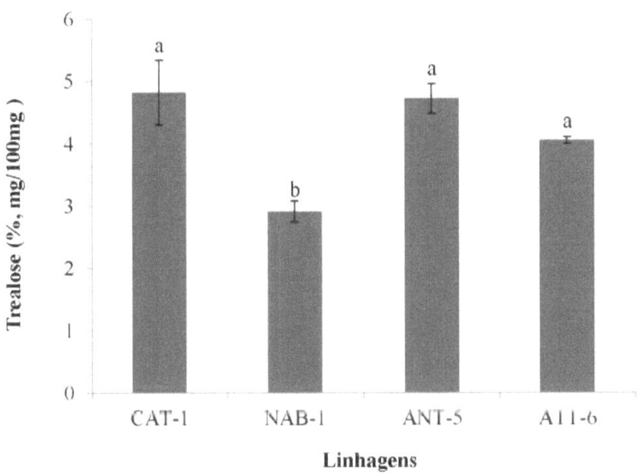

Figure 12. Threhalose (%, mg/100mg dry biomass) and statistical analysis of strains CAT-1, NAB-1, ANT-5 and ATT-6 at the end of the seventh cycle. Equal letters do not differ at 5% significance level, according to Tukey's test.

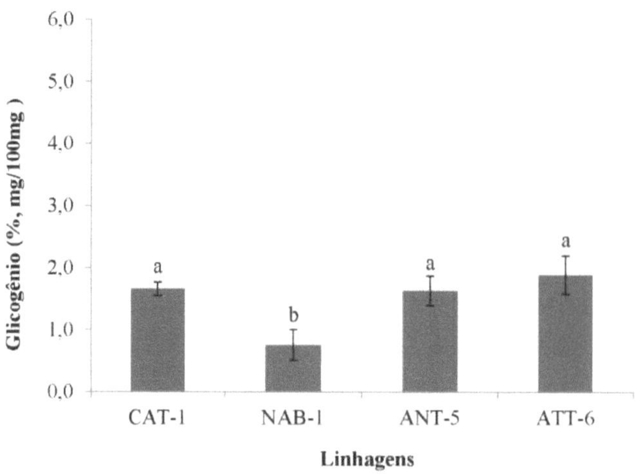

Figure 13. Glycogen (%, mg/100mg dry biomass) and statistical analysis of strains CAT-1, NAB-1, ANT-5 and ATT-6 at the end of the seventh cycle. Equal letters do not differ at 5% significance level, according to Tukey's test.

This study has shown that the ATT-6 strain, with the Msn2 transcription factor (truncated version), could be a potential candidate for the Brazilian process with cell recycling, mainly using fermentations with a high sugar concentration.

In the fourth stage of this work, when the concentrations of sugar cane molasses were drastically increased in the cell cycles, the ATT-6 strain showed greater ethanol production (Figure 7) and CO_2 release (in musts with 18, 22, 28 and 33% ART), and also greater fermentation speed in cycles 1, 2, 4 and 5 (Figure 8), compared to the parental strain. In addition, this strain also had a lower amount of residual ART (%) at the end of cycles 2, 3, 4 and 7, suggesting that ATT-6 utilized more sugar in these cycles (Figure 10). The ATT-6 strain kept its viability stable in the last two cycles, where musts with the highest sugar concentration were used, 35% ART, unlike the parental strain which showed a 10.78% drop in viability in the last cycle analyzed (Figure 9). In the last cycle, ATT-6 did not differ from the parental strain in terms of cell viability (Figure 9). All these results suggest that the over-expression of the *MSN2* gene, in its truncated version, favored the ATT-6 strain under conditions of simultaneous stress, as found in Brazilian distilleries.

70

The ATT-6 strain also stood out in terms of viability in microplate tests using different concentrations of ART in sugarcane molasses must, showing greater viability compared to the parental CAT-1 at 27 and 33% ART (Figure 2). This shows that ATT-6 may have been improved for greater tolerance in a situation of osmotic stress.

When the microscale tests were carried out with different levels of ethanol (in YEPD media), the viability of ATT-6 was the same as the parental CAT-1 at concentrations of 8, 12, 14 and 16% (v/v ethanol). On the other hand, ATT-6 showed lower viability compared to the parental strain when the ethanol content in the YEPD medium was 18% (v/v) (Figure 4). This result suggests that the *MSN2* gene in its truncated version may have influenced the ATT-6 strain to respond mainly to osmotic stress, because when the stress was ethanol, at a very high concentration (18% v/v), this strain proved to be more sensitive compared to the parental strain. This suggests that the experiments designed were able to achieve their purpose of showing the stress limits of each strain after cell recycling, presenting the best fermentation conditions in each experiment.

In short, ATT-6 could be used under conditions of greater osmotic stress, mainly favoring the speed of fermentation. In cycle 4, where must containing 33% ART at 30°C was used, the ATT-6 strain presented the most interesting data for the purpose of this study, i.e. higher ethanol production, fermentation speed, CO release$_2$ and sugar consumption, suggesting that this would be the most suitable must concentration and condition for a possible industrial fermentation using this yeast.

The ANT-5 strain, with the non-truncated version of the *MSN2* gene overexpressed, was similar to the parental CAT-1 in many of the parameters evaluated here, showing no significant differences in terms of growth and viability in the tests with different concentrations of ART (Figure 2) and in most of the parameters of the fermentation experiments with cell recycle analyzed (items 3.3.4 and 3.3.5). However, for the tests with different levels of ethanol (Figure 4), the ANT-5 strain showed lower viability than the parental CAT-1. In this way, it is possible to hypothesize that the level of over-expression of the *MSN2* gene in this version was not enough to result in positive effects in the tests and fermentations with recycles carried out in this work. It is also important to note that the results obtained in this study lead to the inference that the level of overexpression and the form of the *MSN2* gene may influence the strains differently, since different physiological behaviors were observed in the ANT-5 and ATT-6 strains. According to Bücker (2014), overexpression of the *MSN2* gene and *MSN2-T* (truncated version of the gene) play different roles in the general response to stress, mainly by acting in the regulation of genes related to the control of the PKA pathway, inhibiting (ANT-5) or promoting (ATT-6) greater cell proliferation in *S. cerevisiae* yeasts in response to ethanol stress.

The NAB-1 strain, with the over-expressed *TRP1* tryptophan synthesis gene, proved to be less viable in the tests in which different concentrations of ART (%) and ethanol (%, v/v) were tested (first and second stage, Figures 2 and 4), compared to the other three strains. In the cell recycling experiments (items 3.3.4 and 3.3.5), the NAB-1 strain did not differ from its parent in most of the cycles, in most of the parameters analyzed. However, an interesting and specific finding was that it showed significantly greater growth in YEPD medium with 8% ethanol (v/v) (Figure 4A). This corroborates the study by Yoshikawa et al. (2009), in which the importance of the *TRP1* gene in the tolerance of strains to ethanol stress was proven. In this study, the growth behavior of strains with a single deletion in the gene (*TRP1*) under ethanol stress was evaluated and quantified. It was found that the strains with deletions in the "tryptophan metabolism" genes were sensitive to 8% (v/v) ethanol. In another study by Hirasawa et al. (2007), increased expression of genes related to tryptophan biosynthesis gave yeast cells greater tolerance to ethanol stress (15% ethanol v/v), proving in their results that strains with over-expression of these genes could show greater tolerance to ethanol. In the above-mentioned studies on the *TRP1* gene, the cells were exposed to media in which the only stress was caused specifically by ethanol in different concentrations, with ethanol being added to the medium at zero growth time, as was also done in the second stage of this study. It is important to note that in distilleries, the cells withstand the effects of the levels of ethanol produced by them, at the cost of previous cell exposure to high concentrations of ART, i.e. first osmotic stress followed by ethanolic stress. This condition is totally different from the studies in which the efficiency of this gene in increasing tolerance to alcoholic stress has been proven. It is therefore probable that the NAB-1 strain, which over-expresses the *TRP1* gene responsible for tryptophan synthesis, did not show significant differences in the different parameters evaluated here, in the two experiments with cell recycling (where fermentation with a high alcohol content was used), because the process used in the present study was very different from the one carried out in the aforementioned studies, where the efficiency of this gene was proven to increase the yeast's tolerance.

3.4 Conclusions

The present study revealed that the NAB-1 strain, with the *TRP1* tryptophan synthesis gene over-expressed, showed greater growth in YEPD medium containing 8% ethanol (v/v). However, overexpression of the *TRP1* gene did not favor the NAB-1 strain for greater viability when the treatments were concentrations higher than 8% ethanol (v/v) and molasses musts containing 27 and 33% ART. In these treatments, the NAB-1 strain differed from the other strains tested, including the parental CAT-1, with significantly lower viability values. Over-expression of the *TRP1* gene also led to lower accumulation of the reserve carbohydrates trehalose and glycogen. The ANT-5 strain behaved similarly to the CAT-1 strain in most of the parameters analyzed. However, in the microscale growth tests, in YEPD medium with 16 and 18% ethanol, ANT-5 was less viable than ATT-6 and the CAT-1 parent. The results show that over-expression of the *MSN2* gene in the truncated version favored the ATT-6 strain, mainly in relation to higher ethanol production, fermentation speed and higher sugar consumption in the

fermentation experiments with cell recycles, conducted with drastic increases in the must's ART concentration (the last stage of this work). The ATT-6 strain presented the most interesting data for the purpose of this study in the fourth fermentation cycle of the experiment with 7 cycles, where molasses must was used at a concentration of

33% ART at 30°C. ATT-6 also showed significantly higher viability compared to parental CAT-1 in the microscale growth tests where the molasses musts had concentrations of 27 and 33% ART. The level of overexpression of the *MSN2* gene may contribute to different physiological behaviors, since strain ANT-5, with the *MSN2* gene overexpressed in the non-truncated version, differed from strain ATT-6 in the microscale growth tests and in the fermentations with cell recycling. ATT-6 could be used under conditions of greater osmotic stress, favoring fermentation mainly in terms of speed. This study has shown that, among the genetically modified strains studied, the ATT-6 strain has biotechnological potential for fermentation processes using cell recycling and high sugar content in sugar cane molasses must.

3.5 Supplementary Figures

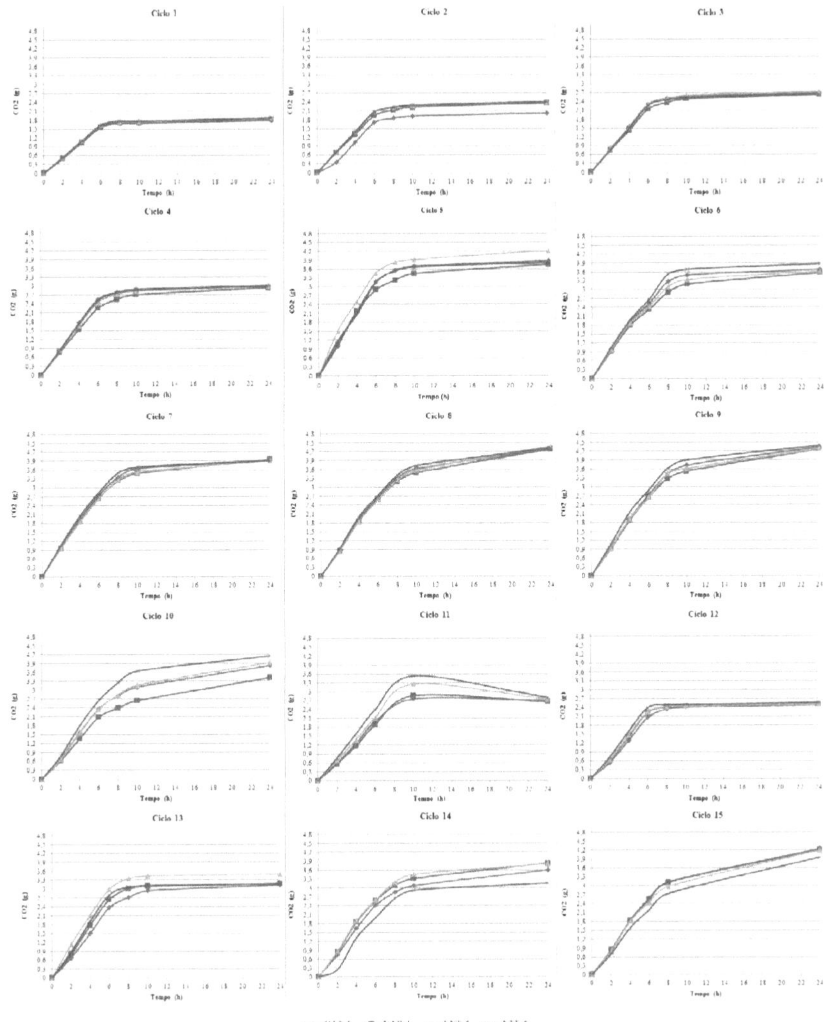

Supplementary Figure 1: Fermentation speed of strains CAT-1, NAB-1, ANT-5 and ATT-6, calculated by CO_2 released (g) over the hours in the fifteen fermentation cycles.

References

Alepuz, P.M.; Jovanovic, A.; Reiser, V.; Ammerer, G. stress-induced map kinase hog1 is part of transcription activation complexes. **Molecular Cell**, v. 7, n. 4, p. 767-777, 2001.

Amorim, H.V.; Basso, L.C.; Alves, D.M.G. **Alcohol production processes**: control and monitoring. Piracicaba, FERMENTEC, 1996. 103p.

Balat, M.; Balat, H. Recent trends in global production and utilization of bio-ethanol fuel. **Applied Energy**, v. 86, n. 11, p. 2273-2282, 2009.

Basso, L.C.; Amorim, H.V.; Oliveira, A.J.; Lopes, M.L. Yeast selection for fuel ethanol in Brazil. **FEMS Yeast Research**, v. 8, n. 7, p. 1155-1163, 2008.

Basso, L.C.; Basso, T.O.; Rocha, S.N. Ethanol production in Brazil: the industrial process and its impact on yeast fermentation. In: SANTOS BERNARDES, M.A. (Ed.). **Biofuel Production - Recent Developments and Prospects**. InTech, 2011, chap. 5, p. 85-100.

Becker, J.U. A method for glycogen determination in whole yeast cells. **Analytical Biochemistry**, v. 86, n. 1, p. 56-64, 1978.

Boinot, F. Melle process of alcoholic fermentation. **The International Sugar Journal**, 1939.

Bücker, A. **Genomic engineering of an industrial strain of *Saccharomyces cerevisiae* to improve ethanol tolerance**. 2014. 111p. Thesis (PhD in Biochemistry) - Federal University of Santa Catarina, Florianópolis, 2014.

Chen, J.C.P.; Chou, C.C. Cane Sugar Handbook - A Manual for Cane Sugar Manufacturers and Their Chemists. In: CHEN, J.C.P.; CHOU,

C.C. (Eds.). **Sugars and non-sugars in sugarcane**. Edison: John Wiley & Sons Inc, 1993, chap. 2, p. 21-30.

Christofoleti-furlan, R.M. Selection of yeasts for high-alcohol fermentation from the biodiversity found in Brazilian distilleries. 2012. 126p. Dissertation (Master's Degree in Agricultural Microbiology) - Luiz de Queiroz College of Agriculture, Piracicaba, 2012.

Cronwright, G.R.; Rohwer, J.M.; Prior, B.A. Metabolic control analysis of glycerol synthesis in *Saccharomyces cerevisiae*. **Applied and Environmental Microbiology**, v. 68, n. 9, p. 4448-4456, 2002.

Debnath, D.; Whistance J.; Thompson, W.; Binfield, J. Complement or substitute: Ethanol's uncertain relationship with gasoline under alternative petroleum price and policy scenarios. **Applied Energy**, v. 191, n.1, p. 385-397, 2017.

Demirbas, A. The social, economic, and environmental importance of biofuels in the future. **Energy Sources, Part B: Economics, Planning, and Policy**, v. 12, n. 1, p. 47-55, 2017.

Estruch, F. Stress-controlled transcription factors, stress-induced genes, and stress tolerance in budding yeast. **FEMS Microbiology Reviews**, v. 24, n.1, p. 469-486, 2000.

Gasch, A.P.; Spellman, P.T.; Kao, C.M.; Carmel-Harel, O.; Eisen, M.B.; Storz, G.; Botstein, D.; Brown, P.O. Genomic expression programs in the response of yeast cells to environmental changes. **Molecular Biology of the Cell**, v. 11, n. 12, p. 4241-4257, 2000.

Godoy, A.; Amorim, H. V.; Lopes M. L.; Oliveira A. J. Continuous and batch fermentation processes: Advantages and disadvantages of these processes in the Brazilian ethanol production. **International sugar journal**, v. 110, n. 1311, p. 175-181, 2008.

Gomar-Alba, M.; Jiménez-Marti, E.; Del-Olmo, M. The *Saccharomyces cerevisiae* Hot1p regulated gene YHR087W (HGI1) has a role in translation upon high glucose concentration stress. **BMC Molecular Biology**, v. 21, n. 13, p. 1-19, 2012.

Gorner, W.; Durchschlag, E., Martinez-Pastor, M.T., Estruch, F.; Ammerer, G.; Hamilton, B.; Ruis, H.; Schüller, C. Nuclear localization of the C2H2 zinc finger protein Msn2p is regulated by stress and protein kinase A activity. **Genes & Development**, v. 12, n. 4, p. 586-597, 1998.

Graves, T.; Narendranath, N.; Power, R. Development of a "stress model" fermentation system for fuel ethanol yeast strains. **Journal of the Institute of Brewing**, v. 113, n. 3, p. 263-271, 2007.

Guo, Z.P.; Zhang, L.; Ding, Z.Y.; Shi, G.Y. Minimization of glycerol synthesis in industrial ethanol yeast without influencing its fermentation performance. **Metabolic Engineering**, v. 13, n. 1, p. 49-59, 2011.

Hersen, P.; McClean, M.N.; Mahadevan, L.; Ramanathan, S. Signal processing by the HOG MAP kinase pathway. **Proceedings of the National Academy of Sciences**, v. 105, n. 20, p. 7165-7170, 2008.

Hirasawa, T; Yoshikawa, K.; Nakakura, Y.; Nagahisa, K.; Furusawa, C.; Katakura, Y.; Shimizu, H.; Shioya, S. Identification of target genes conferring ethanol stress tolerance to *Saccharomyces cerevisiae* based on DNA microarray data analysis. **Journal of Biotechnology**, v. 131, n. 1, p. 34-44, 2007.

Hohmann, S. Osmotic stress signaling and osmoadaptation in yeasts. **Microbiology and Molecular Biology Reviews**, v. 66, n. 2, p. 300-372, 2002.

Iwahashi, H.; Obuchi, K.; Fujii, S.; Komatsu, Y. The correlative evidence suggesting that trehalose stabilizes membrane structure in the yeast *Saccharomyces cerevisiae*. **Cellular and Molecular Biology**, v. 41, n. 6, p. 763-769, 1995.

Kobayashi, N.; Mcentee, K. Evidence for a heat shock transcription factor-independent mechanism for heat shock induction of transcription in *Saccharomyces cerevisiae*. **Proceedings of the National Academy of Sciences**, v. 87, p. 6550-6554, 1991.

Kobayashi, N.; Mcentee, K. Identification of cis and trans components of a novel heat shock stress regulatory pathway in *Saccharomyces cerevisiae*. **Molecular and Cellular Biology**, v. 13, p. 248-256, 1993.

Lewis, J. G. Learmonth, R. P.; Watson, K. Induction of heat, freezing and salt tolerance by heat and salt shock in *Saccharomyces cerevisiae*. **Microbiology**, v. 141, p. 687-694, 1995.

Mager, W.H.; Siderius, M. Novel insights into the osmotic stress response of yeast. **FEMS Yeast Research**, v. 2, n. 3, p. 251-257, 2002.

Mansure, J.J.; Souza, R.C.; Panek, A.D. Trehalose metabolism in *Saccharomyces cerevisiae* during alcoholic fermentation. **Biotechnology Letters**, v. 19, n. 12, p.

1201-1203, 1997.

Marchal, R.; Penet, S.; Solano-Serena, F.; Vandecasteele, J.P. Gasoline and diesel oil biodegradation. **Oil & Gas Science and Technology**, v. 58, n. 4, p. 441-448, 2006.

Marchler, G.; Schüller, C.; Adam, G.; Ruis, H.A. *Saccharomyces cerevisiae* UAS element controlled by protein kinase A activates transcription in response to a variety of stress conditions. **The EMBO Journal**, v. 12, n. 5, p. 1997-2003, 1993.

Munna, M.S.; Humayun, S.; Noor, R. Influence of heat shock and osmotic stresses on the growth and viability of *Saccharomyces cerevisiae* SUBSC01. **BMC Research Notes**, v. 8, p. 1-8, 2015.

Navarro, J.M.; Durand, G. Alcohol fermentation: effect of temperature on ethanol accumulation within yeast cells. **Annals of Microbiology**, v. 129B, n. 2, p. 215-224, 1978.

Nicholls, S.; Straffon, M.; Enjalbert, B.; Nantel, A.; Macaskill, S.; Whiteway, M.; Brown, A.J.P. Msn2- and Msn4-like transcription factors play no obvious roles in the stress responses of the fungal pathogen Candida albicans. **Eukaryot Cell**, v. 3, n. 5, p. 1111-1123, 2004.

Oliveira, A.J.; Gallo, C.R.; Alcarde, V.E.; Godoy, A.; Amorim, H.V. **Methods for microbiological control in alcohol and sugar production**. Piracicaba: FERMENTEC/FEALQ/ESALQ-USP, 1996. 89p.

Park, J.I.; Grant, C.M.; Attfield, P.V.; Dawes, I.W. The freeze-thaw stress response of the yeast *Saccharomyces cerevisiae* is growth phase specific and is controlled by nutritional state via the RAS-cyclic AMP signal transduction pathway. **Applied and Environmental Microbiology**, v. 63, p. 3818-3824, 1997.

Parrou, J.L.; Enjalbert, B.; Plourde, L.; Bauche, A.; Gonzalez, B.; François, J. Dynamic responses of reserve carbohydrate metabolism under carbon and nitrogen limitations in *Saccharomyces cerevisiae*. **Yeast Journal**, v. 15, n. 3, p. 191-203, 1999.

Parrou, J.L.; Teste, M.A.; François, J. Effects of various types of stress on the metabolism of reserve carbohydrates in *Saccharomyces cerevisiae*: genetic evidence for a stress-induced recycling of glycogen and trehalose. **Microbiology**, v. 143, n. 6, p. 1891-1900, 1997.

Pham, T.K.; Wright, T.K. The proteomic response of *Saccharomyces cerevisiae* in very high glucose conditions with amino acid supplementation. **Journal of Proteome Research**, n. 7, p. 4766-4774, 2008.

Puligundla, P.; Smogrovicova, D.; Obulam, V.S.; Ko, S. Very high gravity (VHG) ethanolic brewing and fermentation: a research update. **Journal of Industrial Microbiology and Biotechnology**, v. 38, n. 9, p. 1133-1144, 2011.

Rep, M.; Krantz, M.; Thevelein, J.M.; Hohmann, S. The transcriptional response of *Saccharomyces cerevisiae* to osmotic shock. Hot1p and Msn2p/Msn4p are required for the induction of subsets of high osmolarity glycerol pathway-dependent genes. **Journal of Biological Chemistry**, v. 275, n. 12, p. 8290-8300, 2000.

Sadeh, A.; Movshovich, N.; Volokh, M.; Gheber, L.; Aharoni, A. Fine-tuning of the Msn2/4-mediated yeast stress responses as revealed by systematic deletion of Msn2/4 partners. **Molecular Biology of the Cell**, v. 22, n. 7, p. 3127-3138, 2011.

Saito, H.; Posas, F. Response to Hyperosmotic Stress. **Genetics**, v. 192, n. 2, p. 289-318, 2012.

Sasano, Y.; Watanabe, D.; Ukibe, K.; Inai, T.; Ohtsu, I.; Shimoi, H.; Takagi, H. Overexpression of the yeast transcription activator Msn2 confers furfural resistance and increases the initial fermentation rate in ethanol production. **Journal of Bioscience and Bioengineering**, v. 113, n. 4, p. 451-455, 2012.

Schüller, C.; Brewster, J.L.; Alexander, M.R.; Gustin, M.C.; Ruis, H. The HOG pathway controls osmotic regulation of transcription via the stress response element (STRE) of the *Saccharomyces cerevisiae CTT1* gene. **The EMBO Journal**, v. 13, n.18, p. 4382-4389, 1994.

Singer, M.; Lindquist, S. Thermotolerance in *Saccharomyces cerevisiae*: the Yin and Yang of trehalose. **Trends in Biotechnology**, v. 16, n. 11, p. 460-468, 1998.

Smith, A.; Ward, M.P.; Garrett, S. Yeast PKA represses Msn2p/Msn4p-dependent gene expression to regulate growth, stress response and glycogen accumulation. **The EMBO Journal**, v. 17, n. 13, p. 3556-3564, 1998.

Takagi, H.; Takaoka M, Kawaguchi A, Kubo Y. Effect of L-proline on sake brewing and ethanol stress in *Saccharomyces cerevisiae.* **Applied and Environmental Microbiology**, v. 71, p. 8656-8662, 2005.

Tatebayashi, K.; Yamamoto, K.; Tanaka, K.; Tomida, T.; Maruoka, T.; Kasukawa, E.; Saito, H. Adaptor functions of Cdc42, Ste50, and Sho1 in the yeast osmoregulatory HOG MAPK pathway. **EMBO Journal**, v. 25, n. 13, p. 3033-3044, 2006.

Ter-Schure, E.G.; Van-Riel, N.A.W.; Verrips, C.T. The role of ammonia metabolism in nitrogen catabolite repression in *Saccharomyces cerevisiae.* **FEMS Microbiology**, v. 24, p. 67-83, 2000.

Thomas, K.C.; Hynes, S.H.; Ingledew, W.M. Practical and theoretical considerations in the production of high concentration of alcohol by fermentation. **Process Biochemistry**, v. 31, n. 4, p. 321-331, 1996.

Trevelyan, W.E.; Harrison, J.S. Studies on yeast metabolism. 5. The trehalose content of baker's yeast during anaerobic fermentation. **Biochemical Journal**, v. 62, n. 2, p. 177-183, 1956.

Van-Dijken, J.P.; Scheffers, W.A. Redox balances in the metabolism of sugars by yeasts. **FEMS Microbiology Letters**, v. 32, n. 3-4, p. 199224, 1986.

Waclawovsky, A.J.; Sato, P.M.; Lembke, C.G.; Moore, P.H.; Souza, G.M. Sugarcane for bioenergy production: an assessment of yield and regulation of sucrose content. **Plant Biotechnology Journal**, v. 8, n. 3, p. 263-276, 2010.

Wheals, A.E.; Basso, L.C.; Alves, D.M.; Amorim, H.V. Fuel ethanol after 25 years. **Trends in Biotechnology**, v. 17, n. 12, p. 482-487, 1999.

Yoshikawa, K.; Tanaka, T.; Furusawa, C.; Nagahisa, K.; Hirasawa, T.; Shimizu, H. Comprehensive phenotypic analysis for identification of genes affecting growth under ethanol stress in Saccharomyces cerevisiae. **FEMS Yeast Research**, v. 9, n. 1, p. 32-44, 2009.

Zhao, X.Q.; Bai, F.W. Mechanisms of yeast stress tolerance and its manipulation for efficient fuel ethanol production. **Journal of Biotechnology**, v. 144, n. 1, p. 23-30, 2009.

4 Amino acid supplementation in alcoholic fermentation using CAT-1 in molasses and sugar cane syrup musts

Summary

The production of fuel alcohol from renewable resources has been the subject of considerable interest in recent years. Ethanol production could be improved by the availability of tolerant *Saccharomyces cerevisiae* strains. Studies have investigated the effect of supplements in the substrate to be fermented on cell growth and viability, including yeast extract, ammonium sulphate, urea, etc. Amino acid supplementation can contribute to greater cell tolerance to the different stress conditions faced during the fermentation process. However, the presence of amino acids in substrates used in Brazilian processes has not yet been studied. In this context, this study aimed to evaluate the influence of amino acid supplementation on the growth of the CAT-1 industrial strain under ethanolic and osmotic stress conditions (YNB medium with 10 and 12% v/v ethanol and molasses mash with 15, 20, 25 and 30% ART). In addition, amino acid supplementation was also tested in molasses and sugar cane syrup fermentations, using the CAT-1 strain cell recycle. The results showed that in YNB medium with 12% ethanol (v/v), supplementation with glycine and phenylalanine increased the growth of the strain. On the other hand, supplementation with valine and cysteine in YNB medium with 10 and 12% ethanol (v/v) caused a decrease in the growth of the CAT-1 strain. The presence of cysteine also impaired growth in sugarcane molasses mash with 20% ART. Supplementation with histidine, alanine or lysine increased biomass production compared to the control in molasses fermentations with cell recycling. Tryptophan, asparagine and histidine gave the highest viability values in sugarcane syrup, with asparagine and histidine also promoting greater biomass accumulation in this substrate. The concentration of 200 mg.L^{-1} of amino nitrogen used in the different treatments with amino acids, in YNB medium, and in molasses and cane syrup musts, was effective in differentiating the physiological behavior of the CAT-1 strain. Supplementation with amino acids had different effects on the physiological behavior of CAT-1 yeast, depending on the medium/must supplied. Histidine supplementation favored the CAT-1 strain for greater growth and viability in musts from both molasses and sugar cane syrup. The results revealed that the supplementation of 200 mg.L^{-1} of amino acid N in molasses and sugarcane syrup musts can favor or depreciate the growth and viability of the CAT-1 strain in fermentations simulating Brazilian industrial conditions, with cell recycling. Histidine supplementation proved to be the most promising for increasing the tolerance of the industrial strain under the conditions evaluated here.

Keywords: Histidine; Amino acid supplementation; Alcoholic fermentation; *S. cerevisiae*; Sugarcane syrup and molasses

4.1 Introduction

Understanding the factors that interfere with the ethanol production process is of fundamental importance in order to achieve a fermentation with a high cell survival rate and satisfactory fermentation performance (Basso et al., 2011; Lopes, et al., 2016). An adequate nutritional condition can favor yeast cells, making them able to survive the deleterious effects encountered during fermentation (Tropea et al., 2016). In addition to fermentable sugars and inorganic compounds, nitrogen is an essential component for cells (Albers et al., 1996). In this context, the source of nitrogen available in the substrate during the fermentation process has a direct influence on yeast metabolism and physiology (Kalmokoff & Ingledew et al., 1985; Thomas & Ingledew, 1990). Tolerance, growth, cell multiplication and, consequently, the fermentation yield of the process are examples of factors influenced by the availability of nitrogen in the medium (Thomas & Ingledew, 1990). It is also proposed that in *S. cerevisiae,* nitrogen limitation in fermentations can cause the inactivation of sucrose transporter proteins (Badotti et al., 2008), which is the most abundant sugar in the substrate used in the Brazilian process (Peters, 2006).

The various nitrogen sources used by yeasts are qualitatively referred to as preferred or non-preferred. This classification has been empirically based on two criteria: the first is how well individual compounds support growth when present as the sole nitrogen source, and the second criterion reflects the finding that preferred nitrogen sources generally cause the repression of processes necessary for the utilization of non-preferred nitrogen sources (Magasanik & Kaiser, 2002).

S. cerevisiae yeasts preferentially use nitrogen in the ammoniacal form (NH_4^+) and, from the ammonium ion, synthesize all the amino acids and nitrogenous bases necessary for their growth. Under conditions of ammonium ion deficiency, ammonium N (in the form of urea, for example) or amine N (in the form of amino acids) can be metabolized by other routes of utilization in *S. cerevisiae* (Basso; Amorim, 2001). The nitrogen contained in pure sugarcane juice is mostly available in assimilable form, however, it is insufficient to supply the yeast's cellular nutrition (Chen, 1993; Jeronimo et al., 2008). Supplementation with ammonium sulphate or urea has been demanded by the alcohol industry because they represent more economical sources of N. However, although supplementation with urea has little effect on the pH of the medium, the use of ammonium sulphate can lead to acidity in the external medium, reducing the viability and multiplication of the cells (Amorim et al., 1996). Studying possible sources of N and their respective concentrations suitable for the process is extremely important in order to increase the supplementation options currently available to the industry. Furthermore, some agro-industrial residues contain high concentrations of nitrogen, making them a by-product would be a promising proposal for sustainable development (Woiciechowski et al., 2013). This would require extensive studies on the action of different N sources assimilable by yeast in substrates currently used in different ethanol production processes.

The ability to use amino acids and other nitrogenous compounds requires their internalization and, consequently, yeast cells have multiple permeases to facilitate their transport across the cytoplasmic membrane. The presence of amino acids in the medium induces the expression of several permeases of broad specificity; therefore, amino acids induce their own absorption. This transcriptional response is mediated by the Ssy1-Ptr3-Ssy5 (SPS) sensor located in the cytoplasmic membrane (Ljungdahl,

2009).

Amino acids can be referred to as building blocks, since they perform vital functions for the cell, such as synthesizing enzymes and structural components (Takagi et al., 1997; Morita et al., 2002). According to Hu et al. (2005) the amino acids isoleucine, methionine and phenylalanine can increase the tolerance of *S. cerevisiae* under conditions of ethanol stress. According to the author, the greater tolerance of the cells was due to the incorporation of the supplementary amino acids into the cytoplasmic membrane, which possibly led to a greater capacity to neutralize the fluidization effect exerted by ethanol.

Another study found that a strain overexpressing proline also showed greater viability under conditions of alcoholic stress and, in addition, greater resistance to freezing, desiccation and oxidative stress was observed (Takagi, 2008). Supplementation with glutamic acid has also been shown to benefit *S. cerevisiae* cells, since it promoted greater and faster growth on wheat substrate (Thomas & Ingledew, 1990). Therefore, it is possible to hypothesize that some amino acids may act not only as a source of N, but also as a protective membrane constituent, increasing the cell viability of yeast in adverse situations, such as those encountered in the industrial process of ethanol production. Although there are studies reporting the influence of amino acids on the physiology of *S. ceresisiae,* subjected to different stress conditions, studies highlighting their action individually in fermentations of musts derived from sugar cane are still scarce.

In this context, the aim of this study was to evaluate the supplementation of amino acid nitrogen on the growth and cell viability of CAT-1 industrial yeast in molasses and sugar cane syrup musts.

4.2 Material and Methods

4.2.1 Biological material

The CAT-1 industrial strain was used throughout this study (Basso et al., 2008), in which microscale growth tests and fermentation experiments with cell recycling were carried out in order to evaluate the supplementation of 200 mg.L⁻1 of amino acid N from amino acids.

4.2.2 Micro-scale growth tests

Microscale growth tests were carried out in triplicate on a TECAN spectrophotometer (Optical Density/D.O. 600nm), by incubating 96-well microplates in microaerobic conditions with readings at 2-hour intervals at 30°C for a period of 48 hours.

Shaking was carried out for 5 minutes before each reading. To assemble the microplate, the following was added to each well:

1.1.1 µL of solution of each amino acid (with 1200 mg.L^{-1} of N amine), diluted with water and sterilized in an autoclave at 121 °C for 20 minutes (Table 1);

1.1.2 µL of inoculum pre-grown for 48 hours in YEPD medium (2% D-glucose, 1% yeast extract and 1% bacteriological peptone);

1.1.3 µL of sugarcane molasses mash with different concentrations of ART (%, w/v) and pH 5.0, or 90 µL of YNB medium (0.93% YNB without ammonium sulphate and amino acids, 2.70% D-glucose, 0.1% ammonium sulphate and 0.05% urea) with different levels of ethanol (%, v/v) and pH 5.0.

The concentration of each amino acid was calculated to contribute 200 mg.L^{-1} of amino acid N in 120 µL, this being the final volume of each microplate well (Table 1). The concentrations of urea and ammonium sulphate were calculated to contribute together a total of 200 mg.L^{-1} of amino acid N in a final volume of 120 µL. Amino acids in protein molecules are always L-stereoisomers, so only L-amino acids were used.

Table 1 - Molecular weight and concentration data of the amino acids tested

Amino acids	Molecular weight	mg of amino N.L^{-1} (by millimol.L)$^{-1}$	millimoles.L^{-1} for 1200 mg of amino (1200*14)	mg. L^{-1} for 1200 mg of amino N.L^{-1} (85.71 *molecular weight)	g.L^{-1} for 1200 mg of amino N.L^{-1}
1 Alanine	Wing 89,1	14	85,71	7637,14	7,64
2 Arginine	Arg 210,7	14	85,71	18060,00	18,06
3 Asparagine	Asn 132,1	14	85,71	11322,86	11,32
4 Aspartate or aspartic acid	Asp 133,1	14	85,71	11408,57	11,41
5 Cysteine	Cys 175,6	14	85,71	15051,43	15,05

6	Phenylalanine	Phe	165	14	85,71	14142,86	14,14
7	Glycine or Glycocola	Gly	75,1	14	85,71	6437,14	6,44
8	Glutamate or glutamic acid	Glu	147,1	14	85,71	12608,57	12,61
9	Glutamine	Gln	146	14	85,71	12514,29	12,51
10	Histidine	His	155	14	85,71	13285,71	13,29
11	Isoleucine	Ile	131,2	14	85,71	11245,71	11,25
12	Leucine	Read	131,2	14	85,71	11245,71	11,25
13	Lysine	Lys	182,7	14	85,71	15660,00	15,66
14	Methionine	Met	149,21	14	85,71	12789,43	12,79
15	Proline	Pro	115,13	14	85,71	9868,29	9,87
16	Serina	Be	105,1	14	85,71	9008,57	9,01
17	Tyrosine	Tyr	181,2	14	85,71	15531,43	15,53
18	Threonine	Thr	119,1	14	85,71	10208,57	10,21
19	Tryptophan	Trp	204,2	14	85,71	17502,86	17,50
20	Valine	Val	117,2	14	85,71	10045,71	10,05

1.1.4 Determination of assimilable N in sugarcane molasses

The dosage of assimilable N (aminic N + ammoniacal N) by the yeast was determined in cane molasses must with 10% ART. Initially, the pH of the must was adjusted to 8.0 with NaOH (1.0 mol.L^{-1}), and 10 mL of BaCl2 (1.0 mol.L^{-1}) was added. After 15 minutes, the entire volume was added to a 200 mL volumetric flask and topped up with demineralized water. The entire contents of the flask were filtered through filter paper, discarding the first 10 mL. From the filtrate, 100 mL was transferred to a beaker, and the pH was adjusted again to

8.0 (NaOH, 1.0 mol.L^{-1}) and 25 mL of formaldehyde was added, with the pH also adjusted to 8.0. By titrating NaOH (1.0 mol.L^{-1}) and noting the volume spent, the concentration of assimilable N was calculated (Filipe-Ribeiro & Mendes-Faia, 2007).

1.1.5 Preparation of musts from molasses and sugar cane syrup with amino acid supplementation

The musts from molasses or sugar cane syrup used in the fermentation experiments with cell recycling were prepared by diluting the molasses/syrup with water to obtain the desired percentage (%, w/v) of ART. The diluted musts were centrifuged at 10,000 rpm under 20°C for 20 minutes and distributed in *Erlenmeyer* flasks, which were sterilized at 121°C for 25 minutes. At the same time, a solution of each amino acid containing 2000 mg.L^{-1} of amino acid in 15 mL of distilled water was prepared in 50 mL *erlenmeyer* flasks and sterilized at 121°C for 25 minutes.

In a flow chamber, the entire sterile solution of each L-amino acid was added separately to 350 mL *pet* bottles (previously sterilized with 10% bleach and rinsed with distilled water), followed by 135 mL of molasses mash or syrup. The concentration of amino acids was calculated to be 200 mg.L^{-1} of N amine, giving a final volume of 150 mL of must in each *pet* bottle. The musts were stored in a freezer and thawed daily for use in each cycle.

1.1.6 Propagation of yeast biomass for fermentation experiments

For biomass propagation, 150 µL of the culture previously stored in an ultra-freezer (in a 20% glycerol solution) were transferred to 2 tubes containing 5 mL of YEPD medium (2% D-glucose, 1% yeast extract and 1% bacteriological peptone) and incubated at 28°C for 48 hours. After growth, all the contents of the tubes were transferred to 2 *erlenmeyer* flasks containing 150 mL of sugar cane molasses must from Usina Sao Manoel with 10% ART, pH 4.5, incubated at 30°C for 48 hours. Subsequently, the total contents of each *flask* were transferred to 2 flasks containing 1.2 L of the same must mentioned above. The flasks were incubated at 30°C for 72 hours. The wet biomass was then recovered by centrifugation (4000 rpm) to carry out the first fermentation cycle. The entire procedure for propagating and recovering the wet biomass was repeated for the fermentation experiments with cell recycling.

1.1.7 Fermentation experiments with amino acid supplementation and cell recycling

Two cell-recycling fermentation experiments were carried out using 15 mL Falcon tubes. In the first experiment, with two replicates for each amino acid treatment (200 mg.L^{-1} of N amine), molasses must from the Sâo Manoel Sugar Mill was used. In the second, with three replications for each treatment (200 mg.L^{-1} of amino acid N), sugar cane syrup from Usina Sâo Martinho was used.

In both experiments, for the inoculum of the first fermentation cycle, the biomass of strain CAT-1 from the propagation stage was collected by centrifugation (4000 rpm), so that all the treatments started the first cycle with the same yeast content (g) (0.85 g in the first experiment and 0.76 g in the second). Subsequently, 8 mL of molasses mash was added in the case of the first experiment and cane syrup in the case of the second. The fermentations were conducted at 30°C. The tubes were incubated at 30°C for 8 hours and then kept on the bench at room temperature until the next cycle was carried out the following day. At the end of each fermentation cycle, the yeast biomass was separated from the fermented substrate by centrifugation (4000 rpm) and weighed. Then 8 mL of must was added under the conditions described above. The whole procedure was repeated successively in 7 cycles in the first fermentation experiment, and in 3 cycles in the second. The ART concentration (%, w/v) of the must was gradually increased throughout the first experiment, where the musts had concentrations of 12, 15 and 20% ART. In the second experiment, in which cane syrup was used, the ART concentration was 20%. At the end of all the cycles, the ethanol content (%, v/v) of the fermented wine, the weight of the biomass (g) and cell viability (%) were measured.

1.1.8 Biomass determination

The empty conical-bottomed tubes were weighed before starting the cell-recycling fermentation experiments. At the end of each cycle, after separating the wines, the precipitated wet biomass was weighed (g), and the wet biomass content at the end of each fermentation cycle was determined by the difference in weight of the tube with and without yeast.

1.1.9 Ethanol determination

5 mL of the evaporated wine was transferred into a Kjeldahl micro-distiller (Piracicaba, Brazil). After steam distillation, the condensed material was collected in a 50 mL volumetric flask until the volume was complete. The distilled samples were transferred to an Anton-Paar DMA-48 digital densimeter (Graz, Austria) to estimate the ethanol content (%, v/v). The value generated by the densimeter was multiplied by 10 in order to compensate for the dilution that occurred during the distillation stage.

1.1.10 Determination of cell viability

Cell viability was estimated as a percentage by optical microscopy, under a 40x objective, by differential staining of cells using a solution of erythrosine in a phosphate buffer. Viable cells (unstained) and non-viable cells (stained pink) were counted using a Neubauer chamber. Viability was expressed as the ratio of viable cells to the total number of cells counted (viable and non-viable) (Oliveira et al., 1996).

1.1.11 Statistical analysis

Analysis of variance (ANOVA) and Tukey's mean comparison tests were carried out using the R software. The means for the repetitions of each treatment were considered different at a 5% significance level ($p<0.05$).

4.3 Results and Discussion

4.3.1 Growth tests in YNB medium with ethanol

The main objective of the tests, in which YNB medium containing 10 and 12% ethanol (v/v) was used, was to assess the influence of amino acid supplementation on the growth of the CAT-1 strain when exposed to the levels of ethanol found in the industry. According to Basso et al. (2008) in the ethanol industry, sugar cane juice or molasses, used as substrates in the process, result in production at ethanol concentrations of 8-11% (v/v) within a period of 6-11 hours at 32-35°C.

Growth tests (O.D. 600nm) in YNB medium supplemented with 200 mg.L^{-1} of amino acid N revealed that, when the treatment was 10% ethanol (v/v), over the 48-hour period only aspartic acid (0.760) conferred greater growth than the control (0.713) without amino acid supplementation (control 1) (Figure 1). However, statistical analysis revealed that these final OD values (Asp and control 1) do not represent significant differences. Therefore, in the test with 10% ethanol, no amino acid conferred greater growth compared to control 1, without supplementation.

In this same medium, YNB, without the addition of amino acids and ethanol (control 2), the CAT-1 strain showed an O.D. of 0.881 over 48 hours (Figure 1). With regard to growth impairment, 11 amino acids caused lower growth than control 1: Leu, Pro, His, Ile, Val, Tyr, Met, Lys, Thr, Gln and Cys. Cys supplementation caused significantly less growth than all the other amino acids and controls 1 and 2, suggesting that this amino acid may possibly have been toxic to the yeast under these conditions, since there was almost no growth after 48 hours (O.D. 0.012) (Figure 1).

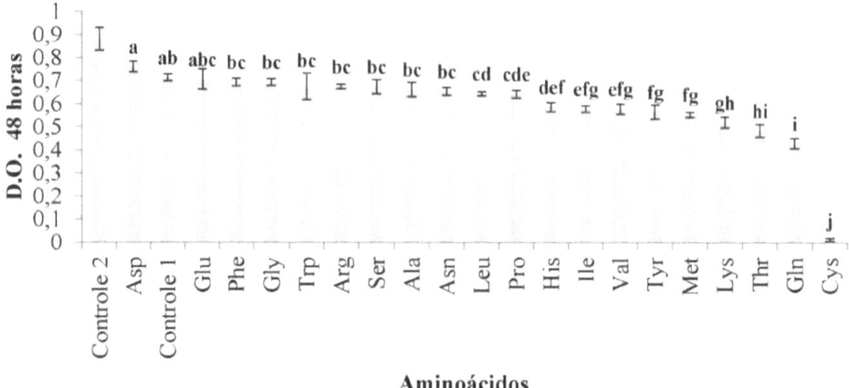

Figure 1. Final O.D. data (600nm) and statistical analysis of the CAT-1 strain at 48 hours in YNB medium with 10% ethanol (v/v) and amino acid supplementation (200 mg.L^{-1} of N amine). Control 1 refers to O.D. in YNB medium with 10% ethanol (v/v) without the addition of amino acids for the same period. Control 2 refers to O.D. in YNB medium without the addition of ethanol and amino acid for the same period. Equal letters do not differ at 5% significance level ($p < 0.05$), according to Tukey's test. Statistical analyses were only calculated for the ethanol treatments.

On the other hand, when the ethanol content was 12% (v/v) in YNB medium, the supplementation of glycine (0.636) and phenylalanine (0.588) significantly increased growth compared to control 1 (0.430) after 48 hours (Figure 2). Glycine is a small achiral amino acid that plays an important role in the carbon metabolism of yeast cells (Piper et al., 2000). Therefore, it plays an important role in supporting cell division (Wu 2009). The three primary glycine pathways are the metabolism and formation of serine via serine hydroxymethyltransferase, oxidative decarboxylation via the glycine cleavage system (glycine synthase), and the formation of nucleotide bases (Sinclair & Dawes, 1995). It has been proposed that phenylalanine is required for optimal growth of *S. cerevisiae* on synthetic media (Hanscho et al., 2012; Crépin et al., 2012). *S. cerevisiae* yeasts synthesize phenylalanine and tyrosine through the intermediates 4-hydroxyphenylpyruvate and phenylpyruvate, respectively. Phenylalanine normally has only three metabolic fates: incorporation into polypeptide chains; production of tyrosine via phenylalanine hydroxylase, which requires tetrahydrobiopterin; and conversion to a fusel alcohol. *S. cerevisiae* degrades aromatic amino acids (tryptophan, phenylalanine and tyrosine) and branched-chain amino acids (valine, leucine and isoleucine) via the Ehrlich pathway. This pathway consists of 3 steps: 1) deamination of the amino acid to the corresponding alpha-hydroxy acid; 2) decarboxylation of the alpha-hydroxy acid resulting in the aldehyde; and 3) reduction of the aldehyde to form the corresponding long-chain alcohol or complex, known as fusel alcohol or fusel oil (Braus, 1991). Fusel oil can be used as a solvent in the cosmetics and pharmaceutical industries, and its main application is to obtain isoamyl alcohol for the synthesis of amyl or isoamyl acetate (a fixative for perfumes) (Güvenç et al., 2007).

One study reported that when the cells of a strain belonging to the *S. cerevisiae* species were exposed to 20% ethanol (v/v) for 9 hours at 30°C, all the cells died. However, 57% of the cells remained viable when the ethanol solution contained three amino acids: isoleucine, methionine and phenylalanine (Hu et al., 2005). Based on analyses of amino acid composition and cytoplasmic membrane fluidity by means of fluorescence anisotropy, using diphenyl hexatriene as a probe, it was revealed that the increased tolerance to ethanol in yeast cells was due to the incorporation of the Supplementary amino acids into the cytoplasmic membranes, which possibly led to a greater ability to neutralize the fluidization effect exerted by ethanol.

Without the addition of ethanol and amino acids, CAT-1 showed an O.D. of 0.924 over the same period. Cistern supplementation caused the same pattern of behavior found in the previous test with 10% ethanol (Figure 1), since with 12% ethanol (v/v), the strain showed an O.D. of 0.003 (Figure 2). This highlights the possibility that the cistern may have been toxic to the strain under these conditions. According to Ono et al. (1999), *S. cerevisiae* strains have various regulatory mechanisms to prevent the overproduction of cisterna and it is likely that cisterna is toxic to the cell because it contains a highly reactive -SH group (organosulfur compound). According to the author, the addition of cisterna at a concentration of 50 μM, i.e. 0.700 mg.L^{-1} of aminic N, caused a delay and reduction in growth, a behavior attributable to the low activity of transport and assimilation of sulphate.

The addition of valine to the YNB medium with 12% ethanol also caused a decrease in growth, and there were no differences between valine (0.147) and cisterna after 48 hours. Derrick & Large (1993) suggest that a significant amount of valine may be metabolized by routes other than the Ehrlich pathway, possibly through the action of branched-chain ketoacid dehydrogenases. These authors suggest that the breakdown of valine requires more energy, and in their study a significantly lower growth yield was obtained from *S. cerevisiae* in medium with valine as the sole source of N. It is worth noting that not all the enzymes involved in the metabolism of branched-chain amino acids in yeast have yet been identified, nor have their interactions been adequately understood.

In short, the addition of the amino acids Lys, His, Thr, Met, Ile, Val and Cys to the YNB medium with 12% ethanol caused a significant decrease in O.D. (48 hours) compared to control 1 (Figure 2). By verifying the data revealed by the tests with 10 and 12% ethanol (v/v), it is possible to suggest that supplementation with the different amino acids can cause different behaviors, increasing or decreasing growth, even under the influence of the concentration of ethanol found in the YNB medium. It is worth noting that the amino acids were not added as the sole source of N in the YNB medium. Therefore, the growth data obtained here was influenced not only by the concentration of ethanol supplied, but also by the ammonium sulphate and urea added to the medium (0.1% and 0.05%, respectively).

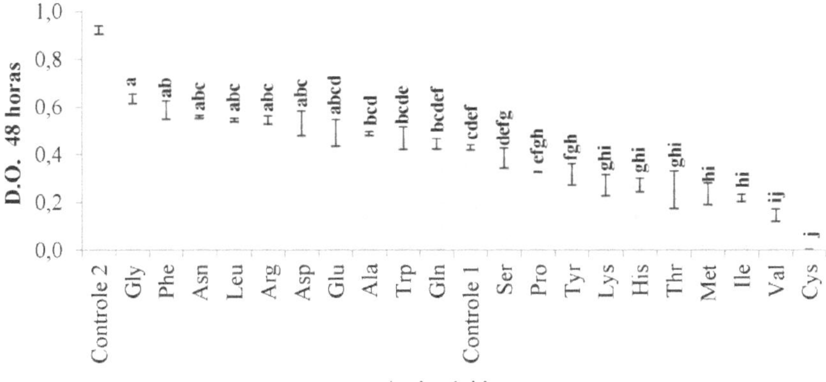

Figure 2. Final O.D. data (600nm) and statistical analysis of strain CAT-1 at 48 hours in YNB medium with 12% ethanol (v/v) and amino acid supplementation (200 mg.L^{-1} of N amine). Control 1 refers to O.D. in YNB medium with 12% ethanol (v/v) without the addition of amino acids for the same period. Control 2 refers to O.D. in YNB medium without the addition of ethanol and amino acid over the same period. Equal letters do not differ at 5% significance level (p<0.05), according to Tukey's test. Statistical analyses were only calculated for the ethanol treatments.

4.3.2 Growth tests and determination of assimilable nitrogen in sugar cane molasses mash

The tests using sugar cane molasses must were carried out to assess both the influence of amino acids on the growth of strain CAT-1 in musts with different concentrations and to select a suitable concentration for subsequent fermentation trials. However, it was necessary to check the dosage of assimilable nitrogen, i.e. aminic N plus ammoniacal N, in order to determine a suitable concentration for amino acid supplementation in the growth tests in molasses. The determination of assimilable N revealed a concentration of 227 mg.L^{-1} in must from molasses from the Sâo Manoel mill, with 10% ART.

According to Amorim et al. (1996), *S. cerevisiae* yeasts have an N content of 8%, based on dry matter, constituting the molecules of amino acids, proteins, enzymes, nucleic acids, purines, primidines, respiratory pigments (cytochromes), vitamins, etc. In YEPD medium (2% D-glucose, 1% yeast extract and 1% bacteriological peptone) the yeast reaches a wet biomass of around 2%, which means that it is possible to obtain 20 g.L^{-1} of wet biomass, or 5 g.L^{-1} of dry biomass (25% of the wet mass).

Therefore, 0.4 g.L^{-1} of N corresponds to 5 g.L^{-1} of dry biomass, assuming that the yeast has an 8% N content, as described above.

In the growth carried out here, we chose to use sugarcane molasses musts in order to approximate the conditions found in Brazilian distilleries. We considered a final content of 3% wet biomass, and that this would require 0.6 g.L^{-1} of N. Knowing that the must from the Sâo Manoel mill (10% ART) has 227 mg.L^{-1} of assimilable N, and that the yeast needs 600 mg.L^{-1} of N to grow 3%, we opted to supplement the molasses musts in the growth tests with 200 mg.L^{-1} of amino acid N from the different amino acids. In this way, it would be possible to concentrate the must, bearing in mind that if we increased the concentration of the must, we would also be increasing the concentration of N. Furthermore, according to Amorim & Leao (2005), the recommended concentration of assimilable N (NH_4^+ and R-NH$_2$) in musts in the fermentation process is 100-300 mg.L^{-1} .

The growth test using sugarcane molasses mash with 15% ART revealed that Val supplementation (1.323) promoted significantly greater growth than the control (O.D. 1.059) after 48 hours of growth. Although the presence of the cistern also caused a decrease in growth (0.909) in the molasses mash, the data showed no statistical difference compared to the control over the same period (48 hours) (Figure 3A). These results show that the medium used can completely alter the effect of supplementation, increasing or decreasing the growth of the strain tested. In other words, the effect on growth promoted by an amino acid can be altered, because this effect can be influenced by the medium. It is important to note that, in *S. cerevisiae,* the metabolic pathways involved in the synthesis and utilization of amino acids often involve the presence of other amino acids, revealing a complex and intricate mechanism (Ljungdahl & Daignan-Fornier, 2012). According to Clement et al. (2013), particular metabolic changes can be triggered according to the nature of the amino acid supplied, in addition to the common response. The presence of some specific amino acids can favor the use of others. For example, in the present study, valine supplementation decreased growth in YNB with 10 and 12% ethanol (Figures

1 and 2), while on the other hand, it promoted greater growth in sugarcane molasses mash with 15% ART (Figure 3A).

Therefore, describing the different effects on growth promoted by amino acids in molasses mash would possibly help to answer which amino acids could have positive or even negative effects on fermentation. Growth is not a primary factor in biotechnological processes where a high density of cells is used, but it is an indication of stress in conditions where there is little oxygen availability. In microaerobic conditions such as those used in this work (microplate assays), the strain may show less growth, or even no growth at all, once the medium is promoting deleterious effects on the cells, as was observed with Cys supplementation.

The growth test in molasses mash with 20% ART revealed that the presence of the cistern significantly reduced the growth (O.D. 0.797) of the CAT-1 strain compared to the control without amino acid supplementation (O.D. 1.001) (Figure 3B). The data obtained from all the growth tests carried out here showed a correlation with regard to Cys supplementation. Both in YNB medium with 10 and 12% ethanol (%, v/v) (Figures 1 and 2) and in molasses mash with 20% ART (Figure 3B), the CAT-1 strain showed significantly lower growth when supplemented with 200 mg.L^{-1} of amino acid N from Cys. These results emphasize the work of Ono et al. (1999), who pointed out that it is likely that cys is toxic to the *S. cerevisiae* cell due to its possession of a highly reactive -SH group.

In must with 20% ART, no supplement other than Cys showed a significant difference compared to the control.

It is possible that the increase in wort concentration led to a higher concentration of total N assimilable by the yeast. This increase may have contributed to the maximum availability of assimilable N required. Therefore, it can be inferred that for this reason there were no significant differences in terms of the CAT-1 strain's greater growth compared to the control without amino acid supplementation. Even higher concentrations of 25 and 30% ART in the molasses mash were evaluated. The results showed no significant differences in CAT-1 growth without and with the supplementation of 200 mg.L⁻1 of amino acid N (Supplementary Figures 1 and 2). Therefore, increasing the concentration of molasses mash, starting at 15% ART, may possibly have less influence on the growth of the CAT-1 strain under the conditions used here. However, in addition to growth, other parameters could be assessed to verify the interference of the presence of amino acids during alcoholic fermentation, and even validate their possible use in the process. The parameters ethanol production and cell viability, for example, are also fundamental for evaluating the effect of the different amino acids on the strain being tested. In addition, the N content of yeast based on dry matter is 8%, but in yeasts during alcoholic fermentation this content is lower, at 5-6% (Amorim et al., 1996). Therefore, in subsequent fermentations with cell recycling, we opted to continue using the concentration of 200 mg.L^{-1} of N from amino acids, in order to evaluate the parameters of cell viability and biomass and ethanol production.

A

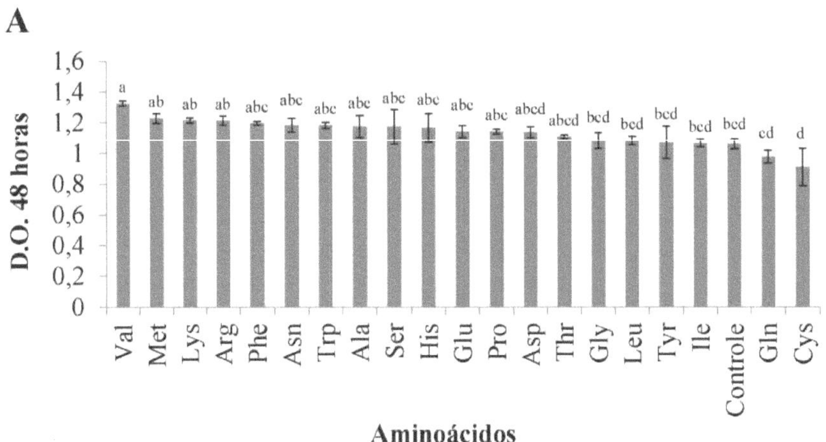

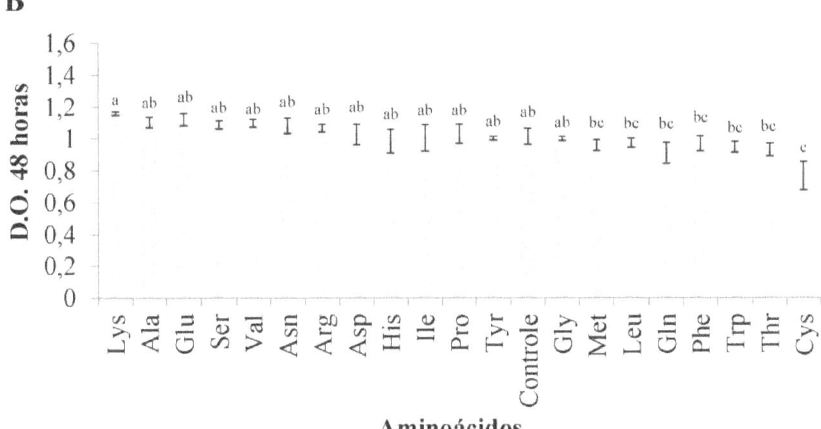

Figure 3 - Final O.D. data (600nm) and statistical analysis of strain CAT-1 at 48 hours in molasses mash with 15% (A) and 20% (B) ART and amino acid supplementation (200 mg.L^{-1} of amino acid N). The control refers to the O.D. without the addition of amino acids over the same period. Equal letters do not differ at 5% significance level (p<0.05)

4.3.3 Fermentation experiments with amino acid supplementation and cell recycling

4.3.3.1 Fermentation in sugar cane molasses must

Based on the O.D. data obtained from the growth tests in molasses mash (item 4.3.2), six amino acids were randomly selected from all those that promoted the greatest growth compared to the control, without amino acid supplementation: Ala, Arg, His, Lys, Ser and Val. Seven consecutive fermentations were carried out in order to verify the influence of the supplementation of the selected amino acids on cell viability (%), biomass production (g) and ethanol production (%, v/v) at the end of each cycle. Amino acid supplementation was 200 mg.L^{-1} of N amine, and for all the treatments with the different amino acids, the initial biomass was 0.85 g (Figure 4).

With regard to wet biomass production (g), there were no significant differences in the first two cycles, since for all treatments, with and without supplementation, the CAT-1 strain had an average of 1.06 g in cycles 1 and 2 (Figure 4). In the subsequent cycles, 3, 4, 5, 6 and 7, supplementation with Ala (1.17, 1.17, 1.19, 1.22 and 1.29 g) and His (1.17, 1.17, 1.17, 1.22 and 1.25 g) led to significantly greater growth compared to the control (1.03, 1.04, 1.04, 1.08 and 1.09 g) (Figure 4). Lys supplementation (1.15 and 1.19 g) also led to greater growth than the control in cycles 5 and 7. The presence of Arg, Ser and Val caused growth significantly equal to the control in all fermentation cycles (Figure 4). The biomass data (g) showed that supplementation with Ala, His, and Lys stimulated greater growth of the CAT-1 strain submitted to sugar cane molasses mash.

Blomqvist et al. (2012) investigated the effect of different amino acids on the number of generations of a *non-Saccharomyces* species (*Dekkera bruxellensis*) in synthetic medium under anaerobic conditions. In this study, corroborating the data obtained here, supplementation with Ala (3.06), His (3.29) and Lys (3.51), also added individually to the medium, significantly increased the number of yeast generations compared to the control (1.04), without supplementation. Dawson (1965) showed that the presence of alanine increased the growth rate of a *Candida utilis* strain in a continuous culture with glucose as the sole carbon source and NH_4 limitation$^+$.

In *S. cerevisiae*, Ala is degraded through transamination to pyruvate by the enzyme alanine aminotransferase. Pyruvate is transported to the mitochondria, where it is oxidatively decarboxylated into carbon dioxide and acetyl-CoA by the pyruvate dehydrogenase complex and then acetyl-CoA enters the citric acid cycle (Schlosser et al., 2004). Under respiratory conditions, alanine aminotransferase plays a central role in the biosynthesis and degradation of arginine. Under fermentative conditions, *Saccharomyces cerevisiae* synthesizes alanine by an alternative pathway, but degrades alanine by the same pathway that is produced during respiration (Garcia-Campusano et al., 2009).

Barton-Wright & Thorne (1949) reported that the presence of the amino acids Lys and His as nitrogenous nutrients for the yeast *S. cerevisiae* is of considerable interest. This is because in their study on the assimilation of 16 individual amino acids in barley wort fermentation, it was found that these amino acids were the first amino acids to be used and, at the end of fermentation, they had been assimilated almost completely. According to the authors, yeasts cannot remove the amino group from Lys and His, and the rapid assimilation, despite the fact that the yeast does not deaminate them, was justified by the hypothesis of intact assimilation of amino acids (Barton-Wright & Thorne, 1949).

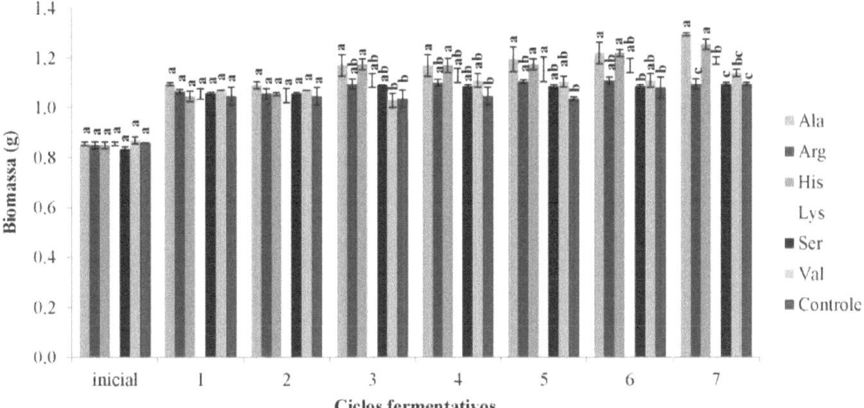

Figure 4: Wet biomass production data (g) and statistical analysis of strain CAT-1 in sugarcane molasses mash with amino acid supplementation (200 mg.L^{-1} of amino acid N) at the end of the 7 fermentation cycles. The control refers to the biomass (g) in sugarcane molasses must without amino acid supplementation. Equal letters within the same cycle do not differ at 5% significance level (p<0.05). The data revealed by the supplementation of each amino acid are arranged in the order shown in the graph legend: Ala, Arg, His, Lys, Ser, Val and Control.

Ethanol production (%, v/v) was the same for all treatments, with and without amino acids. As sucrose was added, from cycle 4 onwards, ethanol also increased. In cycles 1, 2 and 3, where musts with 12% ART were used, for all supplements and the control, average yields were 6.31, 6.86 and 6.55% ethanol (v/v), respectively. In order to increase ethanol stress without increasing molasses concentration, 3% sucrose was added in cycle 4, and 5% sucrose in subsequent cycles (5, 6 and 7). The aim of increasing ethanol production was to reduce the number of viable cells and thus check whether the different amino acid supplements could have an effect on yeast viability. In cycle 4, where must with 15% ART was used, ethanol production was 8.09% (v/v). In cycles 5, 6 and 7, in musts with 20% ART, the different treatments and the control showed average yields of 10.31, 10.43 and 10.46% ethanol (v/v), respectively. The ethanol production data showed that under the conditions in which the experiment was conducted, the different supplements did not alter ethanol production compared to the control.

In terms of viability, the CAT-1 strain started the first cycle with 100% viability for all supplementations and the control. As expected and mentioned above, the addition of sugar from cycle 3 onwards caused a drop in the number of viable cells (Figure 5). There were no significant differences in cell viability in cycles 1, 2 and 3, for all treatments the viability averages were 99.75, 99.53 and 98.80% (Figure 5). Supplementation with 200 mg.L⁻ 1 of N amine from Val showed viability values of 93.14, 86.31, 75.51 and 61.01% in cycles 4, 5, 6 and 7, respectively, which were significantly lower than the control (99.64, 97.29, 90.81 and 78.82%) (Figure 5). Supplementation with Ser also caused a drop in viability from cycle 3 onwards. However, statistical differences were only observed in the last cycle, where this treatment showed a viability of 64.83%, significantly lower than the control's 78.82% (Figure 5). Supplementation with Ala (98.89 and 86.80%), His (99.41 and 99.10%) and Lys (98.18 and 86.94%) promoted the highest viability values in cycles 6 and 7, compared to the control (90.81 and 78.82%) and the other treatments (Figure 5). The treatment with His promoted significantly higher viability compared to the control in cycle 7, and this supplementation promoted the highest viability values compared to all the other treatments and the control in the last two cycles (6 and 7) (Figure 5). In summary, the data obtained from this fermentation experiment showed that the presence of 200 mg.L^{-1} of amino acids from Ala, His and Lys promoted greater accumulation of CAT-1 biomass in sugarcane molasses (Figures 4 and 5). Arg supplementation proved to be the same as the control in terms of viability (%), ethanol production (%, v/v) and wet biomass (g) at the end of all seven cycles (Figures 4 and 5). This suggests that the presence of Arg at a concentration of 200 mg.L^{-1} of N amine, in a medium made from molasses must from the Sao Manoel mill, had no effect on the main fermentation parameters analyzed.

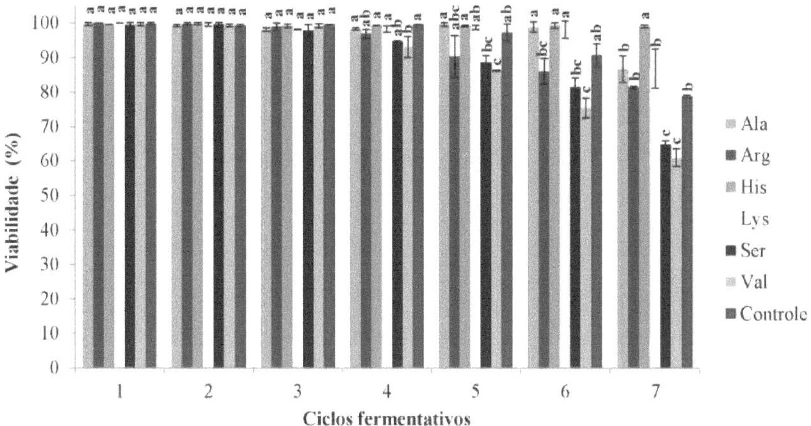

Viability data (%) and statistical analysis of strain CAT-1 in sugarcane molasses must with amino acid supplementation (200 mg.L^{-1} of amino acid N) at the end of the 7 fermentation cycles. Control refers to viability (%) in sugar cane molasses must without amino acid supplementation. Equal letters within the same cycle do not differ at 5% significance level (p<0.05). The data revealed by the supplementation of each amino acid are arranged in the order shown in the graph legend: Ala, Arg, His, Lys, Ser, Val and Control.

According to Jones & Fink (1982), under appropriate conditions, with enough carbon and NH_4^+ available, yeasts can synthesize all the L-amino acids. The families of amino acids derived from a common molecule are easily identifiable and include the glutamate family (glutamate, glutamine, arginine, proline and lysine); the aromatic family (phenylalanine, tyrosine and tryptophan); the serine family (serine, glycine, cisterna and methionine); the aspartate family (aspartate, asparagine, threonine and the sulphur-containing amino acids cisterna and methionine); and the pyruvate family (alanine and the branched amino acids valine, leucine and isoleucine). The histidine and nucleotide biosynthesis pathways are connected. The importance of glutamate and glutamine in the reactions of the central nucleus in nitrogen metabolism is evident, highlighting their involvement in the transamination reactions necessary in the synthesis of each amino acid (Magasanik & Kaiser, 2002; Ljungdahl & Daignan-Fornier, 2012).

With regard to amino acid supplementation in alcoholic fermentation, yeasts tend to accumulate the amino acid supplied and those closely related to it metabolically, and

103

a relatively high concentration of glutamic acid is maintained in the cell regardless of the nitrogen source supplied, reflecting the central role of glutamic acid in nitrogen metabolism (Watson, 1976). In the study by Pham & Wright (2008), supplementation with a complete mixture of amino acids (*Complete Supplement Mixture of Sunrise Science Products*) led to positive cellular responses from *S. cerevisiae,* including reduced latency time and increased cell viability, in media containing a high concentration of sugar (>200 g.L^{-1} of glucose). In addition, ethanol yields increased (0.43 and 0.45 g of ethanol per g of glucose) after supplementation. In this same study, by means of quantitative proteomic analysis, used to understand how *S. cerevisiae* supplemented with amino acids responds to osmotic stress conditions, most of the proteins involved in the glycolysis and pentose phosphate pathways were *upregulated* at high sugar concentrations. The activation of amino acid metabolism was observed at the end of the Lag phase. The relative abundance of the most identified proteins, including aminoacyl-tRNA biosynthesis proteins (transporter RNA linked to an amino acid) and heat shock proteins, remained unchanged in the hours immediately after the addition of glucose. However, the expression of these proteins increased significantly at the end of the Lag phase. In addition, up-regulation of trehalose and glycogen biosynthesis proteins was also observed. These data, combined with the relevant metabolite measurements, show that amino acid supplementation can improve the alcoholic fermentation of media with a high glucose concentration by *S. cerevisiae.*

4.3.3.2 Fermentation in sugar cane syrup must

Sugarcane syrup can also be used in Brazilian distilleries and consists of fewer nutrients and nitrogen sources than molasses. More often, sugarcane juice is concentrated to increase the ART content, and this substrate resembles syrup conveniently diluted for fermentation. For this reason, fermentations were also carried out in syrup must in order to evaluate the influence of amino acid supplementation (200 mg.L^{-1} of amino acid N) on the main fermentation parameters: biomass (g), ethanol (%, v/v) and cell viability (%). In this fermentation trial, significant differences in terms of biomass production and viability were already seen in the different amino acid treatments in just three cycles. These differences were possibly due to the fact that syrup must was used at a concentration of 20% from the first cycle, in order to obtain higher levels of ethanol and, consequently, to reduce the number of viable cells. Based on the data obtained from the previous fermentation experiment, in which sugar cane molasses must was used, the amino acids Ala, Lys and His were selected because they favored the CAT-1 strain in terms of growth and viability (item 4.3.3.1). The amino acids Asn, Glu and Trp were also selected for fermentation in sugarcane syrup, as their presence increased the growth of CAT-1 in microplate assays, both in YNB medium with 12% (v/v) ethanol (Figure 2) and in molasses mash with 15% ART (Figure 3A). In addition, Trp and Glu are amino acids widely cited in the literature for promoting growth and resistance to different types of stress in *S. cerevisiae* (Hirasawa et al., 2007; Yoshikawa, et al., 2009; Godin et al., 2016; Thomas & Ingledew, 1990; Ter-Schure et al., 1998; Wu et al., 2013).

In all treatments, the CAT-1 strain started its first fermentation with 0.76 g of wet biomass, collected from the propagation stage. Already in the first cycle, supplementation with His (1.04 g) led to significantly greater growth than the control (0.92 g) and the other treatments analyzed (Figure 6). The presence of Lys (0.98 g) also promoted significantly greater growth than the control and did not differ from Ala (0.94) and Trp (0.95). In the first cycle, Asn, Glu, Ala and Trp did not differ from the control, without supplementation, in terms of biomass production in sugarcane syrup mash (Figure 6).

In the second and third cycles, the presence of His in the syrup also promoted greater growth (1.14 and 1.18 g) compared to the control (1.06 and 1.06 g) and the other treatments (Figure 6). Although glutamate has been described in the literature as a provider of growth in *S. cerevisiae* yeasts, in syrup must under the conditions described here, it did not cause greater growth than the control in any of the cycles.

In cycle 2, supplementation with Glu, Ala, Asn and Trp produced 0.97, 0.99, 1.03 and 1.03 g of biomass, respectively, which were significantly lower than the control.

Lys supplementation (1.05 g) did not differ from the control in terms of biomass production in the second fermentation cycle (Figure 6).

In the third and final cycle, in addition to His significantly promoting greater growth, the presence of Asn (1.11 g) also showed significantly greater production of wet biomass compared to the control. Supplementation with Glu (0.97 g) showed the lowest biomass accumulation compared to all the other treatments in the last cycle. The amino acids Lys (1.04 g), Ala (1.01) and Trp (1.08) showed similar biomass accumulations to the control in cycle 3 (Figure 6).

In summary, the presence of His and Asn in sugarcane syrup fermentations promoted greater growth of the CAT-1 strain. The biomass production data suggests that although His also favored CAT-1 in molasses mash fermentations, there is no pattern to the effect of the amino acid treatments tested. This is because other amino acids (Ala and Lys) showed higher biomass production in molasses mash compared to the control, and on the other hand, showed similar results to the control in cane syrup mash. In other words, by changing the must, with a different source of minerals and N, the effect of the amino acids may also be modified, altering the physiological profile of the yeast. Nevertheless, His promoted greater growth of the CAT-1 strain in both musts tested here.

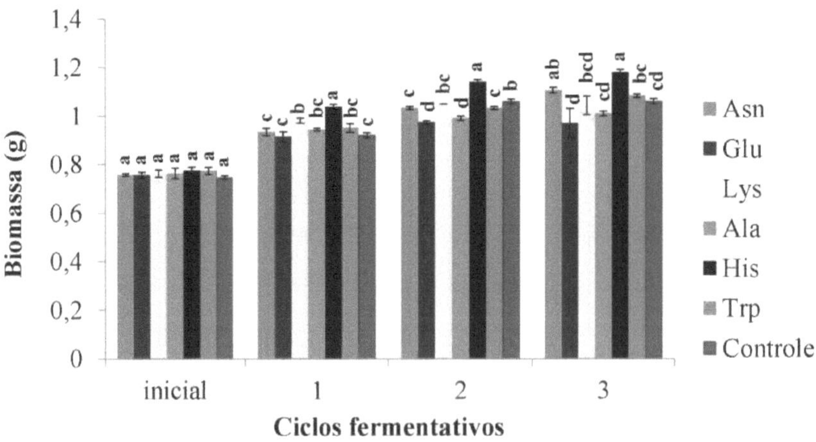

Figure 6. Wet biomass production data (g) and statistical analysis of strain CAT-1 in sugar cane syrup must with amino acid supplementation (200 mg.L^{-1} of amino acid N) at the end of the 3 fermentation

The ethanol production analyses (%, v/v) revealed no significant differences between the different amino acid treatments and the control without supplementation. The production averages, considering all the amino acids and the control, for cycles 1, 2 and 3 are 11.61, 13.38 and 13.71% ethanol (v/v).

With regard to cell viability, all treatments, with and without amino acid, started the first fermentation in syrup with 100% viability. Unlike fermentations in sugarcane molasses mash, supplementation with Ala in syrup decreased CAT-1 cell viability in all three cycles: 90.67% in the first, 78.67% in the second and 77.00% in the third (Figure 7). This data emphasizes the possibility that the amino acids may act in different ways depending on the must/medium used.

In the third cycle, the presence of the amino acids Trp (98.50), Asn (98.50) and His (95.00) showed significantly higher viability values compared to the control (86.00%) (Figure 7). The presence of Glu, even though it showed lower biomass accumulation (Figure 6), showed a viability value of 85.50%, similar to the control in the last cycle (Figure 7). Lys supplementation (88.40%) also did not differ from the control in terms of cell viability.

In syrup must, supplementation with His and Asn favored the strain in terms of growth and viability. In the last cycle, even though the presence of Trp did not increase biomass production, it favored the strain for greater viability in syrup fermentations under the conditions used here.

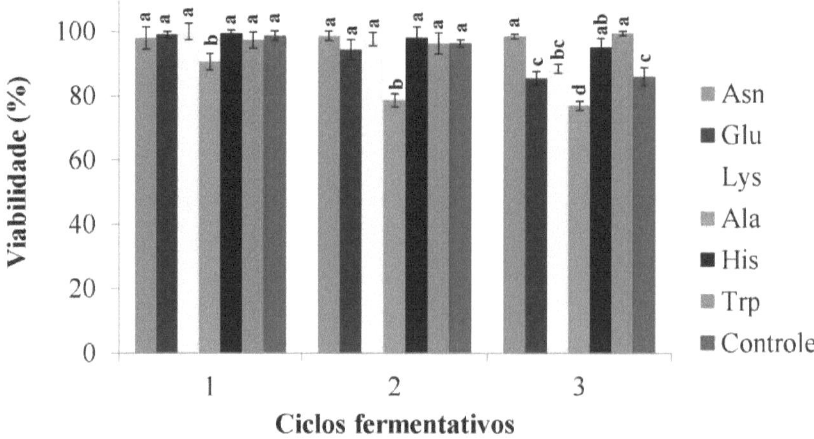

Figure 7. Viability data (%) and statistical analysis of the CAT-1 strain in sugarcane syrup must with amino acid supplementation (200 mg.L^{-1} of amino acid N) at the end of the 3 fermentation cycles. Control refers to viability (%) in

sugar cane molasses mash without amino acid supplementation. Equal letters within the same cycle do not differ at 5% significance level (p<0.05). The data revealed by the supplementation of each amino acid are arranged in the order shown in the graph legend: Asn, Glu, Lys, Ala, His, Trp and Control.

4.4 Conclusions

The concentration of 200 mg.L^{-1} of amino nitrogen used in the different treatments with amino acids, in YNB medium and in molasses and cane syrup musts, was effective in differentiating the physiological behavior of the CAT-1 strain. Based on the data obtained - both from the microplate growth tests and the fermentations with cell recycling - it is possible to state that supplementation with amino acids can have different effects on the physiological behavior of the CAT-1 yeast according to the medium/must used. This is because, by changing the medium, there was a behavioral change in relation to the growth and viability of the strain for most of the amino acids tested. This is possibly due to the fact that amino acids have complex metabolic pathways, comprising an intricate mechanism, often depending on the presence of others for their assimilation/metabolization to occur. Although this behavior was observed for most of the amino acids tested here, supplementation with His showed a pattern of behavior, favoring the CAT-1 strain for greater growth and viability in musts from both molasses and sugar cane syrup. All these data suggest that supplementation with His could be a potential source of N or a protective molecule against the stresses encountered in the fermentation of molasses and sugarcane syrup musts. In summary, the data revealed by this study showed that: in YNB medium with 12% ethanol (v/v), supplementation with Gly and Phe increased the growth of the strain compared to the control. On the other hand, the supplementation of Val and Cys in the YNB medium with 10 and 12% ethanol (v/v) caused a decrease in the growth of the CAT-1 strain. The presence of Cys also reduced the growth of CAT-1 in sugarcane molasses mash with 20% ART in microplate assays. In fermentations with recycle in molasses mash, in addition to His, the amino acids Ala and Lys increased biomass accumulation compared to the control. Supplementation with Val and Ser showed the lowest viability results in molasses mash, in fermentations with recycle. In syrup must, Glu promoted less growth than the control, but showed similar viability. The presence of Trp and Asn, as well as His, also showed the highest viability values in sugarcane syrup. In this same must (syrup), Asn and His also promoted higher biomass production values compared to all the other treatments, with and without amino acids. The results revealed by this study indicate that the supplementation of 200 mg.L^{-1} of amino acids in molasses must and sugarcane syrup can favor or depreciate the growth and viability of the CAT-1 strain in fermentations simulating Brazilian industrial conditions, with cell recycling.

4.5 Additional figures

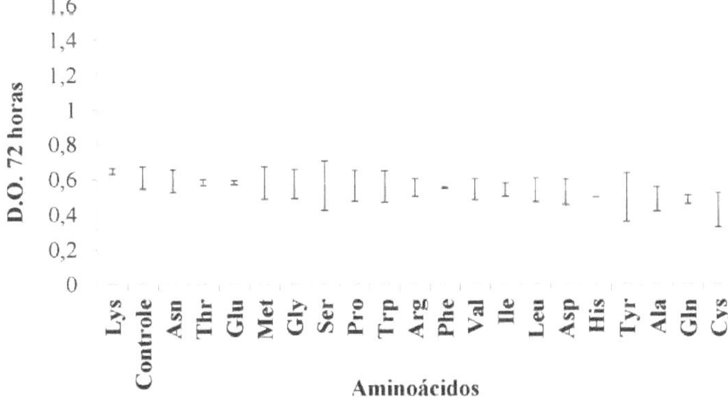

Supplementary figure 1 - Final O.D. data of the CAT-1 strain at 72 hours in molasses mash with 25% ART and amino acid supplementation (200 mg.L⁻1 of amino acid N). The control refers to the O.D. without the addition of amino acids for the same period. Tukey's mean comparison analysis (p<0.05) revealed that there were no significant differences between the different amino acid treatments and the control

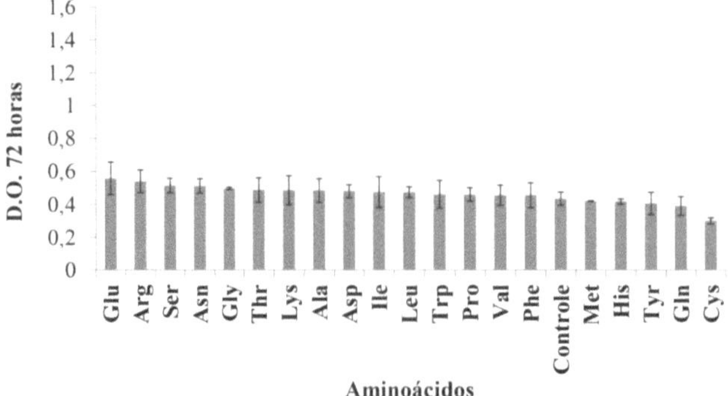

Supplementary figure 2 - Final O.D. data of the CAT-1 strain at 72 hours in molasses mash with 30% ART and amino acid supplementation (200 mg.L⁻¹ of amino acid N). The control refers to the O.D. without the addition of amino acids for the same period. Tukey's mean comparison analysis (p<0.05) revealed that there were no significant differences between the different amino acid treatments and the control

5   FINAL CONSIDERATIONS

Both of the studies carried out here had a common objective: using different approaches, they sought to increase the tolerance of the CAT-1 strain to withstand the stress factors encountered in the Brazilian industrial process.

Compared to the first study, which evaluated *S. cerevisiae* strains with genetic modification, only the ATT-6 strain, with the *MSN2* gene overexpressed (in the truncated version), among the three analyzed, showed favorable parameters for the objective proposed in this study. The level of overexpression or the method of genetic modification may have promoted different physiological behaviors in the strains compared to the parental strain, under the conditions used here. The ATT-6 strain showed greater fermentation speed, sugar consumption and ethanol production in fermentation cycles with drastic increases in the ART concentration of the molasses mash. ATT-6 also showed greater viability than the parental CAT-1 in microscale growth tests in molasses mash with high ART concentrations (27 and 33%, w/v). It is possible to infer that ATT-6 was improved mainly to respond to situations of osmotic stress, since it showed greater or stable viability at the highest ART concentrations in molasses mash tested. The ANT-5 strain was similar to the parental strain in most of the fermentation parameters analyzed, which leads us to hypothesize that the level of overexpression was not enough to have physiological effects in fermentations with cell recycling. Furthermore, ANT-5 was more sensitive to ethanol stress in YEPD medium. The NAB-1 strain, with the *TRP1* gene overexpressed, proved to be more sensitive in the tests in which different concentrations of ART and ethanol were analyzed, compared to the other strains evaluated. In the experiments with cell recycles, NAB-1 was similar to the parent in most parameters, and in YEPD medium with 8% (v/v) ethanol, it showed significantly greater growth. This indicates that tryptophan biosynthesis may have favored the strain to grow more in YEPD medium with this level of ethanol. However, at high ART and ethanol concentrations greater than 8%, the strain (NAB-1) was more sensitive, suggesting that this genetic modification was not efficient for the proposed objective.

In the second study, where the presence of amino acids in the fermentation was evaluated, it was possible to state that their influence on the CAT-1 strain can be completely modified according to the medium provided. In addition, the concentration of 200 mg.L^{-1} of amino acids was efficient in selecting the amino acids that promoted the greatest growth and viability for the yeast. Histidine supplementation proved to be the most promising for increasing the tolerance of the CAT-1 industrial strain in fermentation of molasses mash and sugar cane syrup, using cell recycling.

The results obtained from the experiments in both papers showed that it is possible to increase the tolerance of the CAT-1 industrial strain to the stress factors of fermenting sugarcane musts using the methods employed here - genetic modification and amino acid supplementation.

References

Albers, E.; Larsson, C.; Lidén, G.; Niklasson, C.; Gustafsson, L. Influence of the nitrogen source on *Saccharomyces cerevisiae* anaerobic growth and product formation. **Applied and Environmental Microbiology**, v. 62, n. 9, p. 3187-3195, 1996.

Amorim, H.V.; Basso, L.C.; Alves, D.M.G. **Alcohol production processes**: control and monitoring. Piracicaba, FERMENTEC, 1996. 103p.

Amorim, H.V.; Leao, R.M. **Fermentação alcoólica**: ciência e tecnologia. Piracicaba, FERMENTEC, 2005. 448p.

Badotti, F.; Dàrio, M.G.; Alves, S.L.Jr.; Cordioli, M.L.; Miletti, L.C.; de Araujo, P.S.; Stambuk, B.U. Switching the mode of sucrose utilization by *Saccharomyces cerevisiae*. **Microbial Cell Factories**, v. 7, n. 4, p. 1-11, 2008.

Barton-Wright, E.C.; Thorne, R.S.W. Utilization of amino acids during yeast growth in wort. **Journal of The institute of Brewing**, v. 55, n. 6, p. 383-386, 1949.

Basso, L.C.; Amorim, H.V. Ethanol production. In: Lima, U.A. (Ed.). **Industrial biotechnology: fermentative and enzymatic processes**. Edgard Blucher, 2001, chap. 1, p. 1-43.

Basso, L.C.; Amorim, H.V.; Oliveira, A.J.; Lopes, M.L. Yeast selection for fuel ethanol in Brazil. **FEMS Yeast Research**, v. 8, n. 7, p. 1155-1163, 2008.

Basso, L.C.; Basso, T.O.; Rocha, S.N. Ethanol production in Brazil: the industrial process and its impact on yeast fermentation. In: SANTOS BERNARDES, M.A. (Ed.). **Biofuel Production - Recent Developments and Prospects**. InTech, 2011, chap. 5, p. 85-100.

Blomqvist, J.; Nogué, V.S.; Gorwa-Grauslund, M.; Passoth, V. Physiological requirements for growth and competitiveness of *Dekkera bruxellensis* under oxygen-limited or anaerobic conditions. **Yeast**, v. 29, n. 7, p. 265-274, 2012.

Braus, G.H. Aromatic amino acid biosynthesis in the yeast *Saccharomyces cerevisiae*: a model system for the regulation of a eukaryotic biosynthetic pathway. **Microbiological Reviews**, v. 55, n. 3, p. 349-370, 1991.

Chen, J.C.P. Analysis of juice. In: Chen, J.C.P.; Chou, C.C. (Eds.). **Cane Sugar Handbook - A Manual for Cane Sugar Manufacturers and Their Chemists**. John Wiley & Sons Inc, 1993, chap. 42, p. 931-940.

Clement, T.; Perez, M.; Mouret, J.R.; Sanchez, I.; Sablayrolles, J.M.; Camarasa, C. Metabolic responses of *Saccharomyces cerevisiae* to valine and ammonium pulses during four-stage continuous wine fermentations. **Applied Environmental Microbiology**, v. 79, n. 8, p. 2749-2758, 2013.

Crépin, L.; Nidelet, T.; Sanchez, I.; Dequin, S.; Camarasa, C. Sequential use of nitrogen compounds by *saccharomyces cerevisiae* during wine fermentation: a model based on kinetic and regulation characteristics of nitrogen permeases. **Applied and Environmental Microbiology**, v. 78, n. 22, p. 8102-8111, 2012.

Dawson, P.S.S. The Intracellular amino acid pool of *Candida utilis* during growth in batch and continuous flow cultures. **Biochimica et Biophysica Acta**, v. 3, p. 51-66, 1965.

Derrick, S.; Large, P.J. Activities of the enzymes of the Ehrlich pathway and formation of branched-chain alcohols in *Saccharomyces cerevisiae* and *Candida utilis* grown in continuous culture on valine or ammonium as sole nitrogen source. **Journal of General Microbiology**, v. 139, n. 11, p. 2783-2792, 1993.

Filipe-Ribeiro, L.; Mendes-Faia, A. Validation and comparison of analytical methods used to evaluate the nitrogen status of grape juice. **Food Chemistry**, v. 100, n. 3, p. 1272-1277, 2007.

Garcia-Campusano, F.; Anaya, V.H.; Robledo-Arratia, L.; Quezada, H.; Hernàndez, H.; Riego, L.; Gonzâlez, A. ALT 1-encoded alanine aminotransferase plays a central role in the metabolism of alanine in *Saccharomyces cerevisiae*. **Canadian Journal of Microbiology**, v. 55, n. 4, p. 368-374, 2009.

Godin, S.K.; Lee, A.G.;, Baird J.M.; Herken, B.W.; Bernstein, K.A. Tryptophan biosynthesis is important for resistance to replicative stress in *Saccharomyces cerevisiae*. **Yeast**, v. 35, n. 5, p. 183-189, 2016.

Güvenç, A.; Kapucu, N.; Kapucu,H.; Aydogan, O; Mehmetoglu, Ü. Enzymatic esterification of isoamyl alcohol obtained from fusel oil:

Optimization by response surface methodolgy. **Enzyme and Microbial Technology**, v. 40, n. 4, p. 778-785, 2007.

Hanscho, M.; Ruckerbauer, D.E.; Chauhan, N.; Hofbauer, H.F.; Krahulec, S.; Nidetzky, B.; Kohlwein, S.D.; Zanghellini J.; Natter, K. Nutritional requirements of the BY series of *Saccharomyces cerevisiae* strains for optimum growth. **FEMS Yeast Research**, v. 12, n. 7, p. 796808, 2012.

Hirasawa, T; Yoshikawa, K.; Nakakura, Y.; Nagahisa, K.; Furusawa, C.; Katakura, Y.; Shimizu, H.; Shioya, S. Identification of target genes conferring ethanol stress tolerance to *Saccharomyces cerevisiae* based on DNA microarray data analysis. **Journal of Biotechnology**, v. 131, n. 1, p. 34-44, 2007.

Hu, C.K.; Bai, F.W.; An, L.J. Protein amino acid composition of plasma membranes affects membrane fluidity and thereby ethanol tolerance in a self-flocculating fusant of *Schizosaccharomyces pombe* and *Saccharomyces cerevisiae*. **Sheng Wu Gong**

Cheng Xue Bao, v. 21, n. 5, p. 809-813, 2005.

Jeronimo, E.M.; Souza, E.L.R.; Silva, M.A.; Cruz, J.C.S.; Gava, G.J.C.; Serra, G.E. Soy protein isolate as a nitrogen source in alcoholic fermentation. **Boletim Ceppa**, v.26, n.1, p. 21-28, 2008.

Jones, E.W.; Fink G.R. Regulation of amino acid and nucleotide biosynthesis in yeast. In: Strathern, J.N.; Jones, E.W.; Broach, J.R. (Eds.). **The Molecular Biology of the Yeast *Saccharomyces*: Metabolism and Gene Expression**. Cold Spring Harbor Laboratory Press, 1982, chap. 5, p. 181-299.

Kalmokoff, M.L; Ingledew, W.M. Evaluation of ethanol tolerance in selected *Saccharomyces* strains. **Journal of the American Society of Brewing Chemists**, v. 43, p. 190-196, 1985.

Ljungdahl, P.O. Amino-acid-induced signalling via the SPS-sensing pathway in yeast. **Biochemical Society Transactions**, v. 37, p. 242-247, 2009.

Ljungdahl, P.O.; Daignan-Fornier, B. Regulation of amino acid, nucleotide, and phosphate metabolism in *Saccharomyces cerevisiae*. **Genetics**, v. 190, n. 3, p. 885-929, 2012.

Lopes, M.L.; Paulillo, S.C.L.; Godoy, A.; Cherubin, R.A.; Lorenzi, M.S.; Giometti, F.H.C.; Bernardino, C.D.; Amorim-Neto, H.B.; Amorim, H.V. Ethanol production in Brazil: a bridge between science and industry. **Brazilian Journal of Microbiology**, v. 47S, p. 64-76, 2016.

Magasanik, B.; Kaiser, C.A. Nitrogen regulation in *Saccharomyces cerevisiae*. **Gene**, v. 290, n. 1-2, p. 1-18, 2002.

Morita, Y.; Nakamori, S.; Takagi, H. Effect of proline and arginine metabolism on freezing stress of Saccharomyces cerevisiae. **Journal of Bioscience and Bioengineering**, v. 94, n. 5, p. 390-394, 2002.

Oliveira, A.J.; Gallo, C.R.; Alcarde, V.E.; Godoy, A.; Amorim, H.V. **Methods for microbiological control in alcohol and sugar production**. Piracicaba: FERMENTEC/FEALQ/ESALQ-USP, 1996. 89p.

Ono, B.I.; Hazu, T.; Yoshida, S.; Kawato, T.; Shinoda, S.; Brzvwczy, J.; Paszewski, A. Cysteine biosynthesis in *Saccharomyces cerevisiae*: a new outlook on pathway and regulation. **Yeast**, v. 15, n. 13, p. 1365-1375, 1999.

Peters, D. Carbohydrates for fermentation. **Biotechnology Journal**, v. 1, n. 7-8, p. 806-814, 2006.

Pham, T.K.; Wright, T.K. The proteomic response of *Saccharomyces cerevisiae* in very high glucose conditions with amino acid supplementation. **Journal of Proteome Research**, n. 7, p. 4766-4774, 2008.

Piper, M.; Hong, S.; Ball, G.; Dawes, I. Regulation of the balance of one-carbon metabolism in *Saccharomyces cerevisiae*. **The Journal of Biological Chemistry**, v. 275, n. 40, p. 30987-30995, 2000.

Schlosser, T.; Gatgens, C.; Weber, U.; Stahmann, K.P. Alanine: glyoxylate aminotransferase of *Saccharomyces cerevisiae-encoding* gene AGX1 and metabolic significance. **Yeast**, v. 21, n. 1, p. 63-73, 2004.

Sinclair, D.; Dawes, I. Genetics of the synthesis of serine from glycine and the utilization of glycine as sole nitrogen source by *Saccharomyces cerevisiae*. **Genetics**, v. 140, n. 4, p. 1213-1222, 1995.

Takagi, H.; Iwamoto, F.; Nakamori, S. Isolation of freeze-tolerant laboratory strains of *Saccharomyces cerevisiae* from proline-analogue- resistant mutants. **Applied Microbiology and Biotechnology**, v. 47, n. 4, p. 405-411, 1997.

Takagi, H. Proline as a stress protectant in yeast: physiological functions, metabolic regulations, and biotechnological applications. **Applied Microbiology and Biotechnology**, v. 81, n. 2, p. 211-223, 2008.

Ter-Schure, E.G.; Silljé, H.H.; Vermeulen, E.E.; Kalhorn, J.W.; Verkleij, A.J.; Boonstra, J.; Verrips, C.T. Repression of nitrogen catabolic genes by ammonia and glutamine in nitrogen-limited continuous cultures of *Saccharomyces cerevisiae*. **Microbiology**, v. 144, n. 5, p. 1451-1462, 1998.

Thomas, K.C.; Ingledew, W.M. Fuel alcohol production: effects of free amino nitrogen on fermentation of very-high-gravity wheat mashes. **Applied and Environmental Microbiology**, v. 57, n. 7, p. 2046-2050, 1990.

Tropea, A.; Wilson, D.; Cicero, N.; Potortì, A.G.; La Torre, G.L.; Dugo, G.; Richardson, D.; Waldron, K.W. Development of minimal fermentation media supplementation for ethanol production using two *Saccharomyces cerevisiae* strains. **Natural Product Research**, v. 30, n. 9, p. 1009-1016, 2016.

Watson, T.G. Amino-acid pool composition of *Saccharomyces cerevisiae* as a function of growth rate and amino-acid nitrogen source. **Journal of General Microbiology**, v. 96, n. 2, p. 263-268, 1976.

Woiciechowski, A.L.; Carvalho, J.C.; Spier, M.R.; Soccol, C. Use of agro-industrial residues in food bioprocesses. In: Bicas, J.L.; Maróstica, M.R.; Pastore, G.M. (Eds.). **Food biotechnology**. Atheneu Editora, 2013, chap. 6, p. 143-171.

Wu, G. Amino acids: metabolism, functions, and nutrition. **Amino Acids**, v. 37, n. 1, p. 1-17, 2009.

Wu, Z.; Song, L.; Liu, S.Q.; Huang, D. Independent and additive effects of glutamic acid and methionine on yeast longevity. **PLoS One**, v. 8, n. 11, p. 1-13, 2013.

Yoshikawa, K.; Tanaka, T.; Furusawa, C.; Nagahisa, K.; Hirasawa, T.; Shimizu, H. Comprehensive phenotypic analysis for identification of genes affecting growth under ethanol stress in *Saccharomyces cerevisiae*. **FEMS Yeast Research**, v. 9, n. 1, p. 32-44, 2009.

I want morebooks!

Buy your books fast and straightforward online - at one of world's fastest growing online book stores! Environmentally sound due to Print-on-Demand technologies.

Buy your books online at
www.morebooks.shop

Kaufen Sie Ihre Bücher schnell und unkompliziert online – auf einer der am schnellsten wachsenden Buchhandelsplattformen weltweit! Dank Print-On-Demand umwelt- und ressourcenschonend produziert.

Bücher schneller online kaufen
www.morebooks.shop

Printed by Books on Demand GmbH, Norderstedt / Germany